SpringerBriefs in Applied Sciences and Technology

SpringerBriefs present concise summaries of cutting-edge research and practical applications across a wide spectrum of fields. Featuring compact volumes of 50 to 125 pages, the series covers a range of content from professional to academic.

Typical publications can be:

- A timely report of state-of-the art methods
- An introduction to or a manual for the application of mathematical or computer techniques
- A bridge between new research results, as published in journal articles
- A snapshot of a hot or emerging topic
- An in-depth case study
- A presentation of core concepts that students must understand in order to make independent contributions

SpringerBriefs are characterized by fast, global electronic dissemination, standard publishing contracts, standardized manuscript preparation and formatting guidelines, and expedited production schedules.

On the one hand, **SpringerBriefs in Applied Sciences and Technology** are devoted to the publication of fundamentals and applications within the different classical engineering disciplines as well as in interdisciplinary fields that recently emerged between these areas. On the other hand, as the boundary separating fundamental research and applied technology is more and more dissolving, this series is particularly open to trans-disciplinary topics between fundamental science and engineering.

Indexed by EI-Compendex, SCOPUS and Springerlink.

Manish Kumar Goyal · Vijay Jain · Utkarsh Yadav

Monitoring India's Ramsar Wetlands

Water Quality and Ecosystem Health via Remote Sensing and AI

Manish Kumar Goyal (ORCID)
Department of Civil Engineering
Indian Institute of Technology Indore
Indore, Madhya Pradesh, India

Vijay Jain
Department of Civil Engineering
Indian Institute of Technology Indore
Indore, Madhya Pradesh, India

Utkarsh Yadav
Department of Civil Engineering
Indian Institute of Technology Indore
Indore, Madhya Pradesh, India

ISSN 2191-530X   ISSN 2191-5318 (electronic)
SpringerBriefs in Applied Sciences and Technology
ISBN 978-3-031-96817-4   ISBN 978-3-031-96818-1 (eBook)
https://doi.org/10.1007/978-3-031-96818-1

This Springer imprint is published by the registered company Springer Nature Switzerland AG
The registered company address is: Gewerbestrasse 11, 6330 Cham, Switzerland

If disposing of this product, please recycle the paper.

Preface

Wetlands are ecosystems that lie at the crossing point between land and water, hence significantly playing an important role in providing natural habitats. They sustain a distinctive biodiversity, most being rare or threatened. Anthropogenic influences like pollution, industrial development, urbanization, and global warming have caused a significant impact on wetlands. In order to enhance wetland conservation, the Ramsar Wetlands Agreement was adopted in 1971. It is to encourage the protection and sustainable utilization of wetlands worldwide. This agreement defines wetlands and specifies criteria to list and manage the sites.

In India, 85 wetlands were recognized as Ramsar sites in the year 2024. These sites comprise significant diversity since they range from high-altitude lakes to coastal mangroves. They are broadly classified into three categories: coastal, inland, and human-made. They help in providing ecosystem services like recharging groundwaters, flood regulation, carbon sequestration, etc. Wetlands host millions of people reliant on them for basic resources such as fish, water, and agricultural production. However, these sites face major threats accompanying unsustainable practices, making the conservation of wetlands crucial.

Advanced monitoring technologies, such as remote sensing, artificial intelligence, and cyber-physical systems, therefore, become an important set of tools for wetland management. The book presents the current conditions of these sites with the identification of the major threats using periodic monitoring of them through artificial intelligence and remote sensing.

Keywords Ramsar convention · Wetlands · Water quality · Threats · Site management

Indore, India
Manish Kumar Goyal
Vijay Jain
Utkarsh Yadav

Contents

Chapter 1
India's Ramsar Wetlands: Key Aquatic Ecosystems and Their Significance

Abstract Wetland ecosystems have observed more than a seventy percent decline worldwide in the last 100 years. It significantly impacts the distinctive biodiversity sustained across these sites and the livelihood of the communities across these sites. Presently, these sites observe several threats like human alterations of the catchment, over-exploitation of resources, pollution, and climate change across these sites. In observance of these threats, countries across Europe initiated the Project MAR and Wildfowl meeting to conserve this aquatic ecosystem. It paved the way for the Ramsar Convention in 1971, which later on identified the sites of global importance for the sustenance of biodiversity and human well-being. The significance of Ramsar sites and their site management plan provides the pathway for the conservation of these sites across India. In order to conserve these ecosystems, real-time monitoring of the threats across these sites is required with the integration of emerging technologies. It will assist in developing early warning systems to identify pollution events, invasive species outbreaks, changes in water levels, and wetland health. It will assist in providing precise alerts to the concerned stakeholders for emerging threats and provide an opportunity for rapid intervention to mitigate impacts on wetland ecosystems through timely actions.

Keywords Ramsar wetlands · India · Threats · Water quality · Emerging technologies

1.1 Introduction

Wetlands are comprised of vastly productive aquatic ecosystems that can purify and restore regional water resources [30, 35]. These are considered the transition zones between regional terrestrial and aquatic ecosystems. In accordance with the Ramsar Convention, Article 1.1 wetlands are termed as the "regions comprised marsh, peatland, fen, or water either naturally or artificially with temporary or permanent presence, either static or flowing, fresh brackish or saltwater comprising the regions of marine water with low tide depth not more than six meters" [4, 5, 22]. They are

M. K. Goyal et al., *Monitoring India's Ramsar Wetlands*,
SpringerBriefs in Applied Sciences and Technology,
https://doi.org/10.1007/978-3-031-96818-1_1

1

present worldwide across different regions like riverine floodplains, coastal zones, and urban regions, and act as connecting links in the regional food chain with nature and human beings. Therefore, they significantly impact regional biodiversity, population, ecosystem services, and climate regulation. However, it is estimated that in the last hundred years, these aquatic ecosystems have observed a decline of more than seventy percent worldwide. Considering this loss, the first-time ecosystem-based risk minimization approach was declared as an international theme on World Wetland Day, 2nd February, i.e., "Wetlands and Disaster Risk Reduction" [31]. It is considered since wetlands act as natural buffers for more than ninety percent of water-based disasters. They act as natural sponges by absorbing excess precipitation, supplying water in droughts, and reducing the severity of cyclones, tsunamis, and tornadoes. Therefore, with the rising loss of wetlands and the frequency and magnitude of disasters, the vulnerabilities of the regional population and biodiversity are being enhanced. The rising vulnerabilities are specifically due to a loss of wetland services like food production, regulating hydrological flows, capture, storage, and filtration of water resources, etc. The storage of excess water during disasters enhanced the contribution to restoring groundwater levels for the surrounding aquifers. These wetlands clean the stored water, which is one of the primary functions of swamps and plays an important role in this process. They act as natural filters, like marshes catch debris, sediments, and nutrients from water to feed the wetlands. These aquatic ecosystems comprise distinctive flora and fauna that are tolerant of both land and water conditions. They provide a natal terrain for an abundance of species, namely the migratory birds, fish, amphibians, and insects, where the majority of species breed, feed, and find food in the wetlands [17, 25]. In particular, wetlands are critical to migratory waterfowl, and they are the only places that provide them with a resting place during their long journey. Moreover, they are unique because they are the spawning sites of the fish species that have a great contribution to fisheries [19]. It is estimated that wetland loss can cause annual losses of more than 2.7 and 7.2 trillion US Dollars across swamps, floodplains, tidal marshes, and mangroves, respectively [12]. The United Nations declared the period 2021 to 2030 as a decade for ecosystem restoration in order to benefit the population and nature. This decade aims to cease ecosystem degradation and initiate restoration to increase people's livelihoods, adaptation to climate change, and impacts on biodiversity. The deadline for this decade's meeting to meet the Sustainable Development Goals (SDGs) is considered a crucial point to mitigate the severe impacts of climate change [13, 32]. In India, it is estimated that nearly one-third of natural wetlands have been lost due to urbanization, agriculture, and pollution in the last four decades. The pollution from industrial discharges, agricultural runoff, and untreated sewage creates environmental problems that are considered unhealthy for the wetlands. Overall, it will enhance the crisis for sustainable development, biodiversity protection, food, water, and climate change resilience across the nation [15, 34]. Despite this, India has eighty-five Ramsar-designated sites, which are maximum across the South Asia region. These wetlands are of international importance due to their significance in the conservation of biological diversity and sustaining human well-being [24]. In continuation of anthropogenic

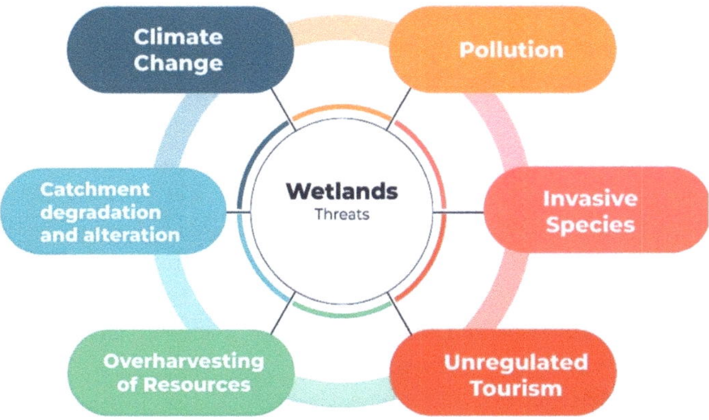

Fig. 1.1 Threats to the wetlands

activities like polluted discharge and encroachment on these sites due to urbanization, agriculture, forestry, etc., changing climatic conditions are threatening these sites significantly. The rising temperature across the states, along with a decline in precipitation amount and the observance of intense precipitation extremes, leads to declines in water content in these sites. These conditions enhance the flooding potential of the sites and minimize the water filtration and storage capacity of these sites. The changing land cover patterns across these sites decrease their natural drainage patterns. It leads to regional catchment degradation and alteration across these sites. Unregulated tourism, excess use of wetland resources, and the presence of invasive species degrade overall wetland health. Overall, the threats faced by the wetlands are shown in Fig. 1.1. The awareness of wetlands protection and their sustainable use has been enhanced through the Ramsar Convention. Though these measures seem to be effective, much needs to be done via correct management and restoration plans for these aquatic ecosystems. Wetlands are presently not only losing species diversity and food and water security but also the ability to respond to environmental change. It is posing a serious risk to the survival of many surrounding communities and biodiversity [19, 26]. Considering this in the subsequent section, we discuss the Ramsar Convention, its criteria, its management plan, and its significance in the Indian context.

1.2 Ramsar Convention

Considering the wetland's importance for the regional biodiversity and population, many small-scale efforts were made to achieve the goal of conservation of the wetlands. However, these efforts were not observed to be sufficient on a larger scale. It led to the formation of the Ramsar Convention in 1971, which covered a broad horizon of conservation. Since then, the Ramsar Convention has brought impactful changes. The process for the Ramsar Convention began in 1960 with Project MAR (Marshes) due to the destruction of huge stretches of marshlands and wetlands across Europe. It has led to a significant decline in the waterfowl region. These concerns were received by the IUCN, which led to the MAR conference and the European Wildfowl meeting in 1961 and 1962, respectively [23].

Consequently, the International Waterfowl and Wetlands Research Bureau (IWRB), today known as Wetlands International, proposed conservation plans for the sites in 1965, followed by the Noordwijk meeting in 1966 for wildfowl conservation. In order to enhance the objectives of Project MAR from Europe to the Middle East, a 1967 wetland conservation support meeting took place in Turkey. Consequently, the Wetlands and Waterbird conferences in 1968 and 1969 at Leningrad (presently St. Petersburg) and Moscow in the Soviet Union happened to finalize the draft for the Wetland Convention. Based on the following meetings, the draft of the Wetland Convention was finalized in 1970 at Knokke, Belgium. Considering this, on 2nd March 1970, the official invitation was sent to all the countries to attend the International Conference on the Conservation of Wildfowl and Wetlands in Ramsar, Iran. Thus, finally, on 2nd February 1971, the Ramsar Convention occurred, and it came into force in 1975 and was signed by seven countries. In 1980, the first conference of the parties (COP) happened at Cagliari, Sardinia, to monitor progress in completing the network of wetlands of international importance. In 1990, at the fourth COP meeting at Montreux, the first list of wetlands with international importance was published. Consequently, on 2nd February 1971, for the first time, World Wetland Day was declared to mark the contribution of the Ramsar Convention [18]. On 30th August 2021, with the Ramsar Convention adopted by 172 countries, the United Nations General assembly adopted the resolution for declaring World Wetland Day as United Nations International Day since 2nd February 2022. It will be used to enhance awareness about the role of wetlands for people and nature worldwide [33]. The overall timeline of the Ramsar Convention has been shown in Fig. 1.2. The Ramsar Convention has established specific criteria for the fulfillment of its mission based on a few articles, which are as follows:

Article 1 defines wetlands broadly as natural or artificial zones of marsh, fen, peatland, or water, whether permanent or temporary, with static or flowing, fresh, brackish, or saltwater, comprising shallow marine regions. The waterfowl are regarded as birds that depend ecologically on these wetlands.

Article 2 states that each participating country must identify the wetlands within its borders to be considered for inclusion in the list of global importance. First, those are especially important as habitats of aquatic plants or animals, and second, those that are of international importance for the breeding, staging, or feeding of waterfowl.

Article 3 mandates that participating parties must create and execute strategies to promote the preservation of specified wetlands and promote their sustainable utilization. Any ecological changes listed in the wetlands because of human activities, such as pollution or technological developments, should be reported.

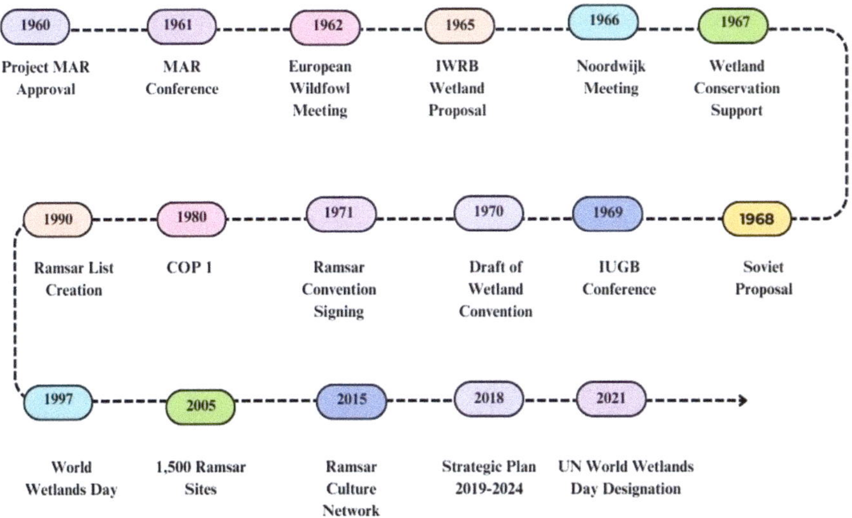

Fig. 1.2 Timeline of Ramsar convention

Article 4 encourages the setting up of nature reserves in wetlands, compensatory measures when listed wetlands are altered, research and exchange of data, and management practices that are designed to further the conservation of waterfowl.

Article 5 provides for international consultation, particularly when wetlands extend over the territories of the parties, especially when there are interdependent water systems concerning conservation measures.

Procedural provisions are laid down in Articles 6–12 regarding the Conference of the Parties to oversee the implementation of the agreement, decision-making processes, and participation of organizations such as IUCN. With amendments made in 1982 and 1987, this agreement is one of the most significant global treaties for the conservation of wetlands. It encourages coordinated national and international policies to safeguard these vital aquatic ecosystems and biodiversity [8]. The criteria for the Ramsar site designation, as decided by these articles, are presented in the subsequent section.

1.2.1   Ramsar Criteria for Site Designation

The criteria for the Ramsar site designation are grouped into two classes, i.e., A and B. Group A comprises criteria 1 for the sites containing representative, rare, or unique wetland types. Criterion 1 identified the wetland as being of global importance if it showcases a natural or nearly natural wetland type that is representative, rare, or unique within the relevant area. Group B comprised the eight criteria from 2 to 9, which centered on sites for the conservation of biological diversity. The details of them are presented as follows:

- **Criterion 2** states that the site aids vulnerable, endangered, or critically endangered species or threatened ecological communities.
- **Criterion 3** states that the site supports the existence of various species that are crucial to maintaining the biological diversity of a particular biogeographic area.
- **Criterion 4** states that the site supports species at an important stage in their life cycles or serves as a refuge while they have unfavorable conditions.
- **Criterion 5** states that the site provides support for at least twenty thousand waterbirds.
- **Criterion 6** states that the site provides habitat for at least 1% of a specific species or subspecies of the waterbird population.
- **Criterion 7** is specific criteria related to fish, stating that the site provides a notable number of native populations that contribute to global biological diversity.
- **Criterion 8** states that the site is a crucial source of food, breeding grounds, nurseries, and migration routes for fish stocks.
- **Criterion 9** states that the site supports 1% of the individuals in a population of a specific species or subspecies of wetland-dependent non-avian animal species.

1.2.2 Site Management Plan

The site management plan is an integrated approach involving the conservation or ecological restoration of the Ramsar sites to attain sustainable use of the wetlands. The key Components of a Ramsar site management plan are as follows:

- **Ecological Assessment**: It assesses the biodiversity, hydrology, and ecosystem function of the wetland to identify key features that need protection and threats that may be ongoing or happening in the future. It informs effective conservation strategy development.
- **Stakeholder Engagement**: It aims to localize this plan through engagements with local communities, government agencies, and indigenous peoples, among others, for its appropriateness to local needs, tapping of traditional knowledge, and community ownership.
- **Identification of Threats**: It aims to identify specific threats to the sites, including pollution, invasive species, and climate change. It assists policymakers in defining ways to minimize such impacts and prevent further deterioration.
- **Conservation and Restoration Actions**: These are the actions required for the maintenance of biodiversity and ecological integrity, including habitat restoration, the exclusion of alien species, and the restoration of natural regimes.
- **Monitoring and Adaptive Management**: It provides an approach for continued monitoring of defined indicators to measure progress. Under this approach, adjustment considering new information or altered conditions promotes resilience to changes caused by natural forces, as well as those from human intervention.

- **Sustainable Use**: It aims to balance the needs of conservation and livelihoods through sustainable endeavors such as eco-tourism, controlled fishing, and sustainable agriculture.
- **Legal and Institutional Framework**: It provides the national and international legal frameworks, besides the institutional arrangements necessary for its execution, legitimacy, and accountability.

The management plan for a Ramsar site is dynamic because it allows periodic review and adjustment to ensure the realization of objectives, as well as maintaining transparency with the Ramsar Secretariat to conserve these essential ecosystems for times to come.

1.2.3 Ramsar Convention in India

India has a total of 15.98 million ha of wetlands, which is nearly 5% of its total geographical area. In February 1982, India gave its approval to the Ramsar agreement. This convention establishes a structure for both national efforts and global collaboration to protect and responsibly utilize wetlands and their resources. As a participant in the convention, India has pledged to uphold the Ramsar Convention principles. India currently has 85 wetlands of global importance, totaling 1,358,066 hectares. Chilika Lake is the oldest among all for inclusion in the list. In August 2024, three new sites were designated as Ramsar wetlands.

Presently, India ranks highest in Asia and third globally in the number of designated sites. Tamil Nadu state has the highest number of wetland sites (18), which are of global importance. It is followed by Uttar Pradesh, which has ten sites. India has distinctive types of wetlands, which are declared Ramsar sites. Their details are presented in a subsequent section.

1.3 Types of Wetlands

The wetlands are broadly classified into three types, i.e., Coastal, Inland, and Manmade wetlands. In this section, we discuss briefly each of these wetlands. However, the detailed assessment will be presented in Chaps. 3 and 4, respectively.

Coastal Wetlands: These sites are present across the transition region of the land and sea. It comprises the characteristics of fresh, saline, and brackish water induced through tidal fluctuations and natural forces. These aquatic ecosystems significantly enhance regional water quality, supporting threatened and endangered species and mitigating the impacts of cyclones, tsunamis, etc. They minimize the intensity and extent of inland flooding. These wetlands serve as nurseries for fish and wildlife for their breeding and feeding regions, with habitats for several species. They have several types, like marine subtidal aquatic beds, intertidal forested wetlands, estuarine waters, and coastal brackish/saline lagoons. India has only sixteen coastal wetlands with Ramsar designation, and they have a major area of 975,202 ha, with a maximum across the Sundarban wetland. The maximum number of these wetlands has an alkaline pH value of more than 7.4 and a salinity range from 0.5 to 30 gL^{-1}. Human influence across these sites, like agriculture, silviculture, urban and port development, with hydrological alteration of excess runoff and exploitation of water resources impact these sites. Natural processes like extreme weather events, erosion, rise in sea level, etc., degrade these ecosystems. A maximum of coastal wetlands is observing the threat of conversion of these sites into open water.

Human-Made Wetlands: These wetlands are formed by human efforts artificially, in replacement to natural wetlands. The main function of these wetlands is flood control and improvement in water quality [7]. However, they are often used to offset the loss of natural habitat by emulating its ecological functions. Their size varies from small ponds and irrigated lands to large water storage reservoirs. India's 41 man-made wetlands include Kolleru Lake and Hirakund Reservoir, which are two of the biggest artificial wetlands in Asia. These sites are generally freshwater reservoirs; however, they are facing challenges in eutrophication because of the high nutrient content discharged into them. These wetlands minimize the effects of urban heat islands across Indian cities and sustain biodiversity in the urban landscape. It will enhance the recreational and eco-tourism activities around these sites. However, these sites are facing threats of pollution through agricultural runoff, industrial effluents, and untreated sewage that may overwhelm the filtration capacity of such wetlands. Invasive species outcompete the native species and alter habitats, reducing biodiversity [3]. Other threats include climate change and urbanization, which further degrade these important ecosystems.

Inland Wetlands: These are either permanently or seasonally water-saturated regions away from the coastal areas [28]. These take different forms, including floodplains, marshes, swamps, and peatlands with water-loving vegetation such as reeds and sedges. Wetlands of this kind are found throughout the world, ranging from freshwater lakes to tundra regions. India has 52 inland wetlands, which are considered important in biodiversity conservation. Their size ranges from small reservoirs to several thousand hectares. Inland wetlands have been losing lots of their area due to dredging, filling, pollution, urban development, and agricultural drainage. Other threats include building dams, river channelization, and over-extraction of water, leading to their degradation and loss.

1.4   Significance of Ramsar Sites and Their Water Quality

These sites have multiple benefits through their ecosystem services for regional populations and environmental sustainability. The surrounding communities depend on these sites for multiple requirements. Being protected wetlands such as Ramsar Sites, these sites have a significant role in sustaining such benefits [10, 36]. Considering this, the water quality needs periodic monitoring across these sites. The importance of these sites and their water quality monitoring are described as follows:

- **Human Food**: These sites are potential sources of food, mainly fish and rice, which are basic nutrition in most regions of the world. Indeed, many Ramsar Sites support fishing communities by using sustainable fishery resources. For example, Chilika Lake in India is a Ramsar Site that alone supports more than 150,000 fisherfolk with livelihoods based on its rich biodiversity. Wetlands also contribute to food security through agriculture-they provide irrigation and land for rice cultivation. The periodic water quality monitoring enhances agricultural production and sustains freshwater sources, biodiversity, and the livelihood of the local community.
- **Eco-tourism Economy**: Ramsar Sites are eco-tourism areas with global value and function as an economic driver for many communities. They sustain rare or even endangered species and have, therefore, become popular sites for bird watchers, nature lovers, and researchers alike. Eco-tourism ensures work and a flow of income that could be reinvested in local conservation work. For example, tourism from around the world flocks to the Sundarbans Wetland, which is a UNESCO World Heritage Site for its rich biodiversity and also houses one of the world's most iconic megafaunas: the Bengal tiger. This economy of tourism is used for the local communities' livelihoods and gives them an incentive to conserve the wild.
- **Water Quality and Purification**: These sites are natural water filters for pollutants and sediments. It ensures good quality water for the human population both inside and around them. Peatlands and mangroves across these sites purify excessive nutrients, heavy metals, and harmful pathogens that reach them. It is used to supply clean water for drinking and agriculture [11]. However, considering the present threats of encroachment and enhanced human activities across these sites, real-time monitoring is required. Water quality parameters like pH, salinity, turbidity, nutrient level, etc., are crucial to be kept under track as they depict the status of the health of wetlands. In this study, we present a discussion on remote sensing-based normalized indices for chlorophyll (NDCI), moisture (NDMI), turbidity (NDTI), and water content (NDWI) to provide the basis for the early indication of threats to the sites. Further, with validated empirical relationships using historical datasets, we can provide real-time monitoring for multiple parameters.
- **Carbon Sequestration**: Wetlands are perfect carbon sinks, holding immense amounts of carbon within their soil and vegetation. Various literature has identified that coastal wetlands, such as mangroves, salt marshes, and seagrasses, offer carbon accumulation rates many times higher than their terrestrial forest counterparts [6]. Hence, designation as Ramsar Sites is considered to help with the offsetting process created by climate change. Sites like the Sundarbans mangrove forests play a vital role in sequestering atmospheric carbon. Thus, they act as important cornerstones in balancing out the rise in greenhouse gas emissions.

- **Habitat Provision**: Ramsar Sites provide diverse habitats to a wide range of diverse species, including some threatened or endemic species. Wetlands like these maintain the processes of bird migration, fish spawning, and the survival of mammals and reptiles. Through wetland health, the Ramsar Sites ensure long-term survival for the various species of global biodiversity. It is, for instance, the habitat of over a million migratory birds during winter and of a variety of fishes, reptiles, and crustaceans that is amazing in its diversity. In order to sustain such benefits of these sites, we will require periodic monitoring of these sites for different threats faced by them using diverse technologies [19, 21].

1.5   Emerging Technologies in Wetland Monitoring

Wetlands are distinctive ecosystems that provide services such as water filtration, carbon sequestration, flood control, and habitats for several species. Consequently, they face significant threats from pollution, climate change, and invasive species. Therefore, in order to manage effectively, such ecosystems require advanced monitoring techniques. Traditional wetland monitoring, relying on field surveying and manual data gathering, is often long, labor-intensive, and of limited scope. The existing techniques used for wetland water quality monitoring in India primarily include a manual collection of water samples. The sample is then tested analytically in the lab to get an idea about the health of the wetland. All these processes take up a lot of time, create a chance for human errors, and can be costly, too. The Central Pollution Control Board (CPCB) has established a water quality monitoring network across the nation's different locations. The present network comprised more than 4484 stations for different water bodies, including wetlands. The water samples for groundwater are collected periodically from them, either monthly, quarterly, or on a yearly basis [20]. In combination with water quality assessment, monitoring of different threats was carried out using the management effectiveness tracking tool (METT) for Indian wetlands. However, this tracking is presently being developed to achieve full monitoring across all the sites with the integration of advanced technologies like artificial intelligence, machine learning, remote sensing, etc. In this context, we discuss the utilization of these advanced techniques for wetland monitoring.

1.5.1   Real-Time Data for Efficiency in Wetland Management and Early Warning Systems

Real-time data becomes an obligatory input in wetland management, especially in monitoring pollution and other threats. Conventional methods used in monitoring often experience delays in collection and processing, hence leading to responses later than necessary for specific changes. Satellite-based remote sensing and Internet-connected sensors are capable of monitoring continuously important indicators such as water quality, hydrology, and land cover. These will provide tools, technologies, and approaches that supply timely and actionable information to identify pollution events, invasive species outbreaks, changes in water levels, and vegetation health. Early warning systems using real-time data can alert managers and stakeholders to emerging threats, thus offering the opportunity for rapid intervention to minimize damage. Indeed, sensors detecting changes in water temperature, pH, or turbidity can provide immediate alerts when pollution incidents occur, while data patterns analyzed automatically may indicate the risk of harmful algal blooms or alien species invasion [9, 29].

Overview of Advanced Technologies: Google Earth Engine, AI, ML, and Cyber-Physical Systems

Recently, several developments in technology-enhanced monitoring and management of wetlands have been initiated worldwide. For example, Google Earth Engine (GEE) grants access to cloud-based processing and analysis of enormous satellite data. GEE is, therefore, useful for wetland monitoring as it allows the development of various datasets that map changes in land cover, changes in water levels, and vegetation health through time, using Landsat and Sentinel satellite imagery. Consequently, exploration of unmanned aerial vehicles (UAVs), artificial intelligence (AI), and machine learning (ML) has also been very effective in monitoring wetlands. They further enhance wetland monitoring through automation in the analysis of big data and recognition of patterns. These technologies can process large volumes of data emanating from satellite imagery, sensors, and other sources to identify trends and anomalies that may signal environmental changes. For instance, ML algorithms can be trained to recognize certain vegetation types from remote sensing imagery or to identify pollution, such as oil spills or algal blooms. Similarly, AI-powered models are used to predict what might change in the future with respect to wetland conditions by looking at historical trends. Thus, it assists managers in making action plans and taking proactive measures to protect such ecosystems. Some of the AI model algorithms include Support Vector Machine (SVM) and Artificial Neural Network (ANN) [14]. Cyber-physical systems (CPS) are an emerging integrated tool for digital and physical processes in environmental monitoring. The data extracted with the help of remote sensing satellites sometimes lacks accuracy. There can be a lot of reasons behind it, including cloud cover and atmospheric refraction [2]. CPS solves the problem to a large extent. It includes the integration of the existing technologies of physical monitoring of water quality with IOT techniques [1]. CPS provides feedback on different environmental parameters in real time. It can measure all types of parameters like pH, salinity, TDS, TSS, etc. These properties assist in monitoring the wetland health and changes in the health of Ramsar sites. It can assist policymakers in identifying the impacts of disasters across these sites with effective strategies for disaster management [27]. It also saves time and cost compared to the physical collection of parameter values. It involves the utilization of field-based water quality sensors connected to the cloud-based server through the Internet to provide efficient real-time monitoring of the sites [16]. As the demand for effective wetland management grows, there's a need for an integrated technological approach that combines modern techniques and methodologies. Advanced real-time monitoring techniques and AI/ML predictive analytics will greatly improve our ability to anticipate changes in wetland ecosystems. This integrated framework will enhance our understanding of wetland dynamics and improve decision-making processes. Therefore, in the next chapter, we will describe the datasets and methodologies for real-time monitoring of threats, water quality, and wetland health across Ramsar sites.

References

1. C. Alexandra, K.A. Daniell, J. Guillaume, C. Saraswat, H.R. Feldman, Cyber-physical systems in water management and governance. Curr. Opin. Environ. Sustain. **62**, 101290 (2023). https://doi.org/10.1016/j.cosust.2023.101290
2. B. Cao, R. Deng, S. Zhu, Universal algorithm for water depth refraction correction in through-water stereo remote sensing. Int. J. Appl. Earth Obs. Geoinf. **91**, 102108 (2020). https://doi.org/10.1016/j.jag.2020.102108
3. Y. Chen, X. Wang, M. Li, L. Liu, C. Xiang, H. Li, Y. Sun, T. Wang, X. Guo, Impact of trace elements on invasive plants: attenuated competitiveness yet sustained dominance over native counterparts. Sci. Total Environ. **927**, 172292 (2024). https://doi.org/10.1016/j.scitotenv.2024.172292
4. L.M. Cowardin, *Classification of Wetlands and Deepwater Habitats of the United States* (US Department of the Interior Fish and Wildlife Service Office of Biological Services, 1979)
5. C.S.S. Ferreira, M. Kašanin-Grubin, M.K. Solomun, S. Sushkova, T. Minkina, W. Zhao, Z. Kalantari, Wetlands as nature-based solutions for water management in different environments. Curr. Opin. Environ. Sci. Health **33**, 100476 (2023). https://doi.org/10.1016/j.coesh.2023.100476
6. K. Gallant, P. Withey, D. Risk, G.C. van Kooten, L. Spafford, Measurement and economic valuation of carbon sequestration in Nova Scotian wetlands. Ecol. Econ. **171**, 106619 (2020). https://doi.org/10.1016/j.ecolecon.2020.106619
7. R.M. Gersberg, S.R. Lyon, R. Brenner, B.V. Elkins, Integrated wastewater treatment using artificial wetlands: a gravel marsh case study, in *Constructed Wetlands for Wastewater Treatment* (CRC Press, 2020), pp. 145–152
8. M.K. Goyal, S. Kumar, A. Gupta, *AI for Water Conservation* (2024a), pp. 17–29. https://doi.org/10.1007/978-3-031-72014-7_2
9. M.K. Goyal, S. Kumar, A. Gupta, *AI Framework for Future Water* (2024b), pp. 55–67. https://doi.org/10.1007/978-3-031-72014-7_5
10. M.K. Goyal, S. Kumar, A. Gupta, *Basics of AI for Water Management* (2024c), pp. 1–16. https://doi.org/10.1007/978-3-031-72014-7_1
11. F. Huang, Y. Huang, J. Jia, Z. Li, J. Xu, S. Ni, Y. Xiao, Research and engineering application of bypass combined artificial wetlands system to improve river water quality. J. Water Process Eng. **48**, 102905 (2022). https://doi.org/10.1016/j.jwpe.2022.102905
12. IUCN, *Wetlands and Disaster Risk Reduction* (IUCN, 2017). https://iucn.org/news/water/201701/world-wetlands-day-strengthening-resilience-and-collaboration-reduce-disaster-risk
13. V. Jain, K.S. Rautela, M.K. Goyal, Ecological restoration: an overview of science and policy regime. Ecosyst. Restor. Towards Sustain. Resilient Dev. 1–27 (2023)
14. R. Khatun, S. Talukdar, S. Pal, T.K. Saha, S. Mahato, S. Debanshi, I. Mandal, Integrating remote sensing with swarm intelligence and artificial intelligence for modelling wetland habitat vulnerability in pursuance of damming. Eco. Inform. **64**, 101349 (2021). https://doi.org/10.1016/j.ecoinf.2021.101349
15. R.T. Kingsford, A. Basset, L. Jackson, Wetlands: conservation's poor cousins. Aquat. Conserv. Mar. Freshwat. Ecosyst. **26**(5), 892–916 (2016). https://doi.org/10.1002/aqc.2709

16. Y. Lei, Y. Rao, J. Wu, C.-H. Lin, BIM based cyber-physical systems for intelligent disaster prevention. J. Ind. Inf. Integr. **20**, 100171 (2020). https://doi.org/10.1016/j.jii.2020.100171

17. L. Lorrain-Soligon, F. Robin, X. Bertin, M. Jankovic, P. Rousseau, V. Lelong, F. Brischoux, Long-term trends of salinity in coastal wetlands: effects of climate, extreme weather events, and sea water level. Environ. Res. **237**, 116937 (2023). https://doi.org/10.1016/j.envres.2023.116937

18. G.V.T. Matthews, *The Ramsar Convention on Wetlands: Its History and Development* (2013). https://leap.unep.org/sites/default/files/2020-09/Matthews-history.pdf

19. F. McDuie, A.A. Lorenz, R.C. Klinger, C.T. Overton, C.L. Feldheim, J.T. Ackerman, M.L. Casazza, Informing wetland management with waterfowl movement and sanctuary use responses to human-induced disturbance. J. Environ. Manage. **297**, 113170 (2021). https://doi.org/10.1016/j.jenvman.2021.113170

20. MoEFCC, Analysis of water quality of rivers by Central Pollution Control Board (2023). Retrieved 19 Nov 2024, from https://pib.gov.in/PressReleseDetailm.aspx?PRID=1941065®=3&lang=1

21. A. Piralizefrehei, M. Kolahi, J. Fisher, Ecological-environmental challenges and restoration of aquatic ecosystems of the Middle-Eastern. Sci. Rep. **12**(1), 17229 (2022). https://doi.org/10.1038/s41598-022-21465-0

22. Ramsar, *What are Wetlands?* (2007). https://www.ramsar.org/sites/default/files/documents/library/info2007-01-e.pdf

23. Ramsar, *1962–1970* (Ramsar, 2024a). https://www.ramsar.org/1962-1970

24. Ramsar. *Ramsar Sites Information Service* (Ramsar, 2024b). https://rsis.ramsar.org/ris-search/

25. C. Sivaperuman, C. Venkatraman, Coastal and marine bird communities of India. In *Marine Faunal Diversity in India* (Elsevier, 2015), pp. 261–281

26. S. Singh, M.K. Goyal, A review of India's water policy and implementation toward a sustainable future. J. Water Clim. Change **16**(2), 493–510 (2025). https://doi.org/10.2166/wcc.2025.560

27. S. Singh, M.K. Goyal, E. Saikumar, Assessing climate vulnerability of Ramsar wetlands through CMIP6 projections. Water Resour. Manage **38**(4), 1381–1395 (2024). https://doi.org/10.1007/s11269-023-03726-3

28. E. Stirling, R.W. Fitzpatrick, L.M. Mosley, Drought effects on wet soils in inland wetlands and peatlands. Earth Sci. Rev. **210**, 103387 (2020). https://doi.org/10.1016/j.earscirev.2020.103387

29. TNSWA, *Tamil Nadu Wetlands Mission* (2025). Retrieved 16 Mar 2025, from https://www.tnswa.org/pallikaranai-marsh

30. UNEP, *Wetlands Limit Impact of Floods, Drought, Cyclones* (UNEP, 2017a). https://www.unep.org/news-and-stories/story/wetlands-limit-impact-floods-drought-cyclones

31. UNEP, *Wetlands Limit Impact of Floods, Drought, Cyclones* (UNEP, 2017b)

32. United Nations, *The United Nations Decade on Ecosystem Restoration* (2021). https://wedocs.unep.org/bitstream/handle/20.500.11822/31813/ERDStrat.pdf?sequence=1&is Allowed=y

33. United Nations, *World Wetlands Day 2022* (UN, 2022). https://www.un.org/en/observances/world-wetlands-day/background

34. WISA, India lost one-third of its natural wetlands in four decades, reveals study. Hindustan Times (2020). https://www.hindustantimes.com/cities/india-lost-one-third-of-its-natural-wetlands-from-1970-to-2014/story-QmhTehlWAcep0cSHdbzufI.html

35. H. Wu, R. Wang, P. Yan, S. Wu, Z. Chen, Y. Zhao, C. Cheng, Z. Hu, L. Zhuang, Z. Guo, H. Xie, J. Zhang, Constructed wetlands for pollution control. Nat. Rev. Earth Environ. **4**(4), 218–234 (2023). https://doi.org/10.1038/s43017-023-00395-z

36. Y. Ye, H. Qiu, Environmental and social benefits, and their coupling coordination in urban wetland parks. Urban For. Urban Green. **60**, 127043 (2021). https://doi.org/10.1016/j.ufug.2021.127043

Chapter 2
Approaches and Techniques for Wetland Analysis

Abstract In this chapter, we present the Government of India's actions for the conservation of the Ramsar wetlands across the nation under different policy frameworks. Consequently, we assessed the socio-economic importance of these sites in regional community development and biodiversity conservation. Further, we describe the various datasets and methodologies used for continuous monitoring of the wetlands, such as water quality, health, and threats, using the Google Earth Engine, artificial intelligence, machine learning techniques, etc. Lastly, we present the criteria and statistical assessment techniques for the assessment of water quality and wetland health, along with the limitations of the datasets and methodologies described.

Keywords Geographic overview · Datasets · Monitoring techniques · Limitations

2.1 Introduction

India is one of the richest, biodiverse wetland regions, and many of these sites have cultural and religious significance. The wetlands are conserved sites under various national policies like the Indian Forest Act (1927), the Indian Wildlife Protection Act (1972), and the Forest Conservation Act (1980), respectively [1]. Further, the government of India integrated wetland conservation under different national frameworks. For example, the Indian National Environment Policy (2006) enhances policy and priority action requirements for wetlands. The National Biodiversity Action Plan (2008) integrated wetlands management in completing national biodiversity conservation objectives. Consequently, the Convention on Biological Diversity objectives, along with the Aichi biodiversity targets (2010), considered the wise use of wetland

---

The original version of the chapter has been revised. A correction to this chapter can be found at
https://doi.org/10.1007/978-3-031-96818-1_6

M. K. Goyal et al., *Monitoring India's Ramsar Wetlands*,
SpringerBriefs in Applied Sciences and Technology,
https://doi.org/10.1007/978-3-031-96818-1_2

resources. National Water Mission (2011), under the National Climate Action Plan, considers wetland conservation as a strategic response to mitigating the impacts of climate change. Considering the importance of wetlands and their ecosystem services, the government of India initiated a National Action Plan for the Conservation of Aquatic Ecosystems (2013) specifically for wetlands and lakes. Subsequently, with the observance of the declining migratory birds' populations, the National Action Plan for the Conservation of migratory birds and their habitats along the Central Asian Flyway was initiated in 2018. Further, the National Wildlife Action Plan (2017–2031) identified the conservation of inland.

aquatic ecosystems as one of the seventeen priority areas [2, 3]. Considering these policy frameworks for enhancing regional sustainability, India is enhancing its Ramsar sites network with the maximum number of sites in the South Asian region. In the period from 1982 to 2013, India had only 26 sites with Ramsar designation. However, in a short period of ten years between 2014 and 2024, the Government of India enhanced its commitment to wetland conservation by the addition of 59 new Ramsar sites [1]. India presently has eighty-five Ramsar sites with an area of 13,67,749 ha. spread from the northern Himalayan region to southern coastal areas and from western arid ecosystems to eastern floodplains. The natural wetlands among them comprised Himalayan lakes, wetlands across floodplains of major river basins, saline and temporary wetlands across arid and semi-arid regions, coastal wetlands like lagoons, backwaters, mangrove swamps, coral reefs, etc. [4]. However, with these diverse wetlands' presence across the nation and high dependence on rising food and water security, there is significant pressure on wetlands, specifically urban wetlands. Urban wetlands have shown a maximum decline with the city's expansion. It leads to a decline in their services, such as disaster resilience, green infrastructure, community livelihoods, enhancement in air and water quality, etc. Considering their importance, the Ramsar Convention initiated the Wetland City Accreditation scheme. It will assist urban regions worldwide in conserving, restoring, and sensibly utilizing urban wetlands. Considering these three Indian cities, i.e., Indore, Bhopal, and Udaipur, were nominated in the year 2024 for wetland city accreditation [5, 6]. Further, in this chapter, we discuss India's progress and the regions lacking in monitoring datasets and methodologies for enhancing regional wetland health and water quality.

2.1.1 India's Ramsar Sites Importance and Conservation Progress

India's Ramsar sites are of deep importance to international conservation efforts due to the endemic biodiversity they support and the critical ecosystem services they provide. The Ramsar sites provide habitat for globally threatened species like the Siberian crane, Indian skimmer, and Gangetic dolphin, as well as turtles and saltwater crocodiles. These wetlands have acted as important stopovers for flyways such as the Central Asian Flyway, with millions of birds resting and feeding on them while breeding there [7, 8]. The Chilika Lake is one such wetland that supports more than a million migratory waterfowl during winter. It thus remains one of the most significant bird congregation sites in Asia. Similarly, the Sundarbans wetlands, the largest tidal halophytic mangrove forest worldwide, depict rich biodiversity with habitat for the endangered Royal Bengal Tiger and saltwater crocodile.

Beyond their biological significance, such wetlands have an immense role to play in hydrological stability by enabling the recharge of groundwater, flood control, sediment entrapment, and shoreline stabilization [9, 10]. These ecosystems are significant carbon sinks, thus moderating the effect of climate change by sequestering humongous quantities of carbon dioxide. For example, mangroves across the Sundarbans wetlands alone store millions of tons of carbon annually. Thereby, these Ramsar sites would denote India's commitment to wetlands worldwide that are of international importance, with their crucial services transcending international borders. It is through such protection that the biodiversity of the world is preserved, the resilience of ecosystems maintained, and the livelihoods of millions of people sustained [11].

Presently, the government of India has initiated the "Wetlands Rejuvenation Programme," with a selection of 130 wetlands across the nation. This program collects baseline information, prepares wetland health cards, collaborates with "Wetland Mitras" to enable stakeholders to take participatory action in wetland conservation, and prepares the site management plan for precise actions. The nation also notified the Wetlands (conservation and management) rules (2017) within the provisions of the Environment Protection Act, 1986. Several workshops have been organized to strengthen the capacity of wetland managers to manage their work effectively. However, despite this, wetland managers have observed several challenges in implementing the Ramsar Convention, such as the absence of effective mechanisms for sharing best practices. Consequently, the convention does not link the financial mechanism in order to support the management of sites through promoting their values, falling in the biodiversity-centric criteria for site designation, and not addressing the daily challenges faced by the wetland managers [12]. These wetlands have economic and cultural values as they enhance community development and livelihoods across these sites, which is presented in a subsequent section. The details of all the Ramsar sites across India are shown in Fig. 2.1.

2.1.2 Socioeconomic Context and Community Interactions with Wetlands

Wetlands have significant socio-economic importance for millions of individuals as these sites support them in their agricultural production, disaster management, tourism, fishing, etc. [13]. For example, Chilika Lake comprises thirty percent of the fishing community in its surroundings, and Sundarbans provide honey, fish, and fuelwood to local communities. Further, these sites are inextricably linked to cultural or religious traditions as well. For example, Loktak Lake in Manipur supports

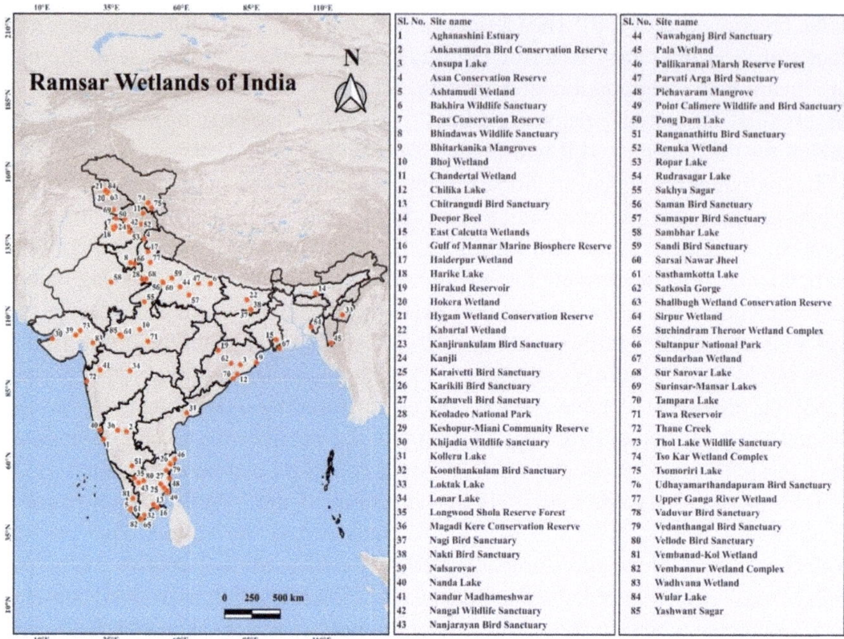

Fig. 2.1 Location map of Ramsar wetlands of India

Indigenous Meitei people who are dependent on the resources for traditional fishing and agriculture [14]. However, wetlands have been subject to intense pressure due to human population, overfishing, agricultural runoff, industrial waste, etc. It is impacting the biodiversity and livelihoods of the community in its surroundings. Balancing ecological health with socio-economic needs and conservation efforts to protect such ecosystems requires the participation of local communities in decision-making processes, sustainable tourism promotion, and livelihood diversification [15]. The following are successful examples: Loktak Lake Community-based Conservation and Mangrove Management in the Sundarbans, where conservation has successfully coincided with local interests [16, 17]. Consequently, we will now discuss the datasets and methodologies used for monitoring threats, specifically water quality and wetland health.

2.2 Datasets and Methodologies

There are different types of datasets and methodologies available for the monitoring of wetlands with the integration of remote sensing, artificial intelligence, machine learning, cyber-physical systems, etc. These datasets give us an idea about the current condition of the wetland's health and the water quality of the sites.

2.2.1 Datasets for Wetland Monitoring

Wetland monitoring aims to identify the threats and their severity assessment by consolidating various environmental data. It assists us in presenting the health of the wetland ecosystem and the changing land use dynamics. The foremost important dataset in wetland health monitoring is climatic conditions. The changing climatic conditions can be assessed and monitored using the regional meteorological department and satellite open-access datasets of temperature, precipitation, humidity, water levels across floodplains and the sea, etc. Climatic variables datasets can be accessed using satellites like MERRA-2, GPM, ERA-5, CHIRPS, etc. Further, the influence of climatic oscillations like El Nino southern oscillations, Indian Ocean Dipole, and Atlantic multidecadal oscillations can be assessed on the regional climatic conditions. The climatic datasets assist us in enhancing disaster management efficiency for precipitation and temperature extremes [18]. It will be used to assess the relationship between climatic conditions and changes in regional water quality and ecosystem health. Climatic variables are the key drivers in initiating the changes in wetland ecosystems and tracking the surrounding environmental conditions [19, 20]. When assessing the climatic variables, we can monitor the water quality parameters like pH, turbidity, dissolved oxygen, salinity, pollutants, etc., for the wetlands. It is primarily done by using the field-based sensor datasets connected with the cloud server for real-time monitoring, machine learning and artificial intelligence data analytics, citizen participation, drone mapping, and remote sensing assessment through validated empirical relationships. In order to utilize machine learning and artificial intelligence, refined field-based big datasets are necessary for efficient prediction of the water quality parameters in different seasons. The remote sensing assessment can be done effectively with the imagery of Landsat and Sentinel satellites. The normalized indices computation, like chlorophyll (NDCI), turbidity (NDTI), moisture (NDMI), and water (NDWI) content index, provides the basic water quality assessment. The details of these indices are presented in Table 2.1. These parameters are important for assessing the overall health status of the aquatic ecosystem. However, the human influence across these sites needs to be assessed. Human influence is computed through the changing land use across these sites, such as agricultural activities and their runoff, built-up encroachment, mining activities, etc. The land use dynamics across the wetland sites can be effectively monitored periodically using the remote sensing assessment through Landsat and Sentinel satellite imagery. The integration of these datasets will subsequently offer a holistic perspective on the impacts and risks arising from changing climatic conditions, quality of water, and land use across Ramsar sites.

The chlorophyll index is used to identify eutrophication across water bodies due to agricultural runoff and industrial or domestic discharge. Consequently, the moisture index provides information on the vegetation conditions present in the water bodies. The turbidity index indicates the suspended pollutants present in the water bodies, while the water index provides the freshwater content available in the sites. The entire assessment can be carried out using the Google Earth engine for time series

Table 2.1 Description of water quality indices

Water quality indices	Description	Formula
NDCI	Normalized Difference Chlorophyll Index	$\frac{\text{Red Edge 1} - \text{Red}}{\text{Red Edge} + \text{Red}}$
NDTI	Normalized Difference Turbidity Index	$\frac{\text{Red} - \text{Green}}{\text{Red} + \text{Green}}$
NDMI	Normalized Difference Moisture Index	$\frac{\text{NIR} - \text{SWIR}}{\text{NIR} + \text{SWIR}}$
NDWI	Normalized Difference Water Index	$\frac{\text{Green} - \text{NIR}}{\text{Green} + \text{NIR}}$

satellite datasets. Further, in order to assess the wetland's vulnerability precisely to climate change, water quality, and human influence, wetland health card details need to be computed. The wetland health card is comprised of several datasets, which are presented below in Table 2.2. In order to compute the risk for this aquatic ecosystem, we have to compute the sites' exposure based on the number of ecosystem services they provide to the regional biodiversity and population. The six prominent ecosystem services considered as per the Ramsar convention are as follows: nutrient cycling, biodiversity, pollination, food for humans, freshwater, and recreation with tourism activities.

Therefore, considering the hazards associated with climate change, water quality, and human influence, along with vulnerable wetland health and exposure to ecosystem services, we can compute the risk for the aquatic ecosystem of the wetlands across the nation using Eq. 2.1.

$$\text{Risk (R)} = H * V * E \tag{2.1}$$

In combination with the remote sensing assessment, multiple data sources, like historical data, including government reports, such as Ramsar information sheets, provide information regarding hazards for wetland health and water quality. The integration of satellite data, field surveys, and historical records provides the best way to come up with a holistic and accurate assessment of environmental factors. It assists us in assessing the present wetland conditions and, over the long term, facilitates a clearer understanding of the threats and their associated risks to the Ramsar sites.

2.2.2 Datasets Collection Periods and Methodology

In order to enhance the site management and monitoring of the water quality and wetland health across the Indian Ramsar sites, we require real-time monitoring using field-based sensor datasets. In combination with this, we required high spatiotemporal resolution satellite image assessment using Sentinel-2 or Landsat-8 with validated empirical relationships for different water quality parameters using Google Earth Engine (GEE). Once we import a high spatiotemporal image through selected satellite imagery, we can filter images with appropriate cloud cover, preferably less than 10%. This cloud masking removes all the images with a cloud cover of more than 10%. In order to remove noise and remove outliers, we can perform techniques like moving average smoothing, median filtering, machine learning regression, etc., and join them using the appropriate techniques. Once we get a noise-free time series of images, we have to fill the cloudy pixels using unmasked pixels either through spatial kernel or temporal interpolation techniques. After the gap-filling of images, we have to create a regular time series of images in order to assess the water quality datasets for each time step. The interpolation techniques are used to fill the empty images in the timestep from adjacent images to provide a regular time series. Considering these regular time series images, we can make image composites for specific land cover patterns and normalized water quality indices like NDCI, NDTI, etc., for daily monitoring of Ramsar wetland sites [10, 21]. This type of automated model in GEE assists in monitoring daily spatial changes in the water quality, along with the development of empirical relationships with different climatic conditions. In association with this, artificial intelligence and machine learning techniques [22, 23] are effective in assessing wetland risk [24]. The study by Ghosh and Das [25] addresses the critical issue of wetland conversion by proposing a novel methodology for risk assessment, specifically focusing on the East Kolkata Wetland (EKW). As traditional approaches often rely on flawed assumptions that can lead to inaccuracies. In contrast, this study integrates advanced machine learning techniques, namely, Random Forest (RF) and Support Vector Machine (SVM), with environmental modeling to enhance predictive accuracy. While SVM demonstrates slightly superior performance, both models effectively identify vulnerable areas within the EKW, particularly near the Kolkata megacity, where urban pressures increase conversion risks. The resulting risk maps are essential tools for policymakers and conservationists, guiding targeted interventions in high-risk zones. Moreover, this methodology establishes a framework applicable to other wetlands facing similar threats, thereby contributing to global wetland conservation efforts. The key outcomes of this study include the demonstrated effectiveness of RF and SVM in predicting conversion risk, the identification of urban proximity as a significant risk factor, and the provision of reliable risk maps for informed conservation strategies. Further, developing a high-resolution water quality model is an emerging field, and an appropriate solution using an artificial intelligence approach for determining the various water quality parameters. For example, the DynQual water quality model was built by Jones et al. [26] using the DynWat framework, which used the global water temperature model and energy-water balance. It is used to simulate daily water temperature and three water quality constituents globally from 1980 to 2019, i.e., total dissolved solids (mg/L), fecal coliform count, and biological organic matter demand (BOD), which are key for social and environmental assessment. This model is comprised of socio-economic and climate forcing as primary components. The socio-economic forcing datasets considered the gross and net water demand for the domestic, industrial, and livestock populations and their mean effluent concentrations. The climate forcing datasets comprised datasets like precipitation, temperature, potential evapotranspiration, solar radiation, sunlight hours, vapor pressure, and cloud cover percentage. In combination with these, hydrological datasets like surface runoff through urban regions, industrial centers, and irrigation fields were considered. Considering these modeled outcomes, we can identify the pollutant hotspots across the globe and determine the trend in deteriorating water quality with climatic and anthropogenic activities [27]. In addition to this threat monitoring of wetland sites, we required an assessment of wetland health considering the water quality datasets and government policies for enhancing their resilience with efficient conservation plans. The integration of these approaches in monitoring threats to wetlands and their water quality assists in developing effective conservation strategies and facilitating targeted interventions to protect ecosystem health.

2.3 Research Methodology: Approaches and Techniques for Wetland Analysis

In the subsequent chapters, we will present the assessment of the water quality indices to a greater extent, including land cover patterns in wetland surroundings and ecosystem health. It employs a multi-faceted approach to assess wetland ecosystems, focusing on water quality and land cover dynamics. Data collection involves remote sensing, historical records, and government reports, integrating various tools and techniques for a comprehensive analysis.

2.3.1 Methodical Schematic Approach

See Fig. 2.2.

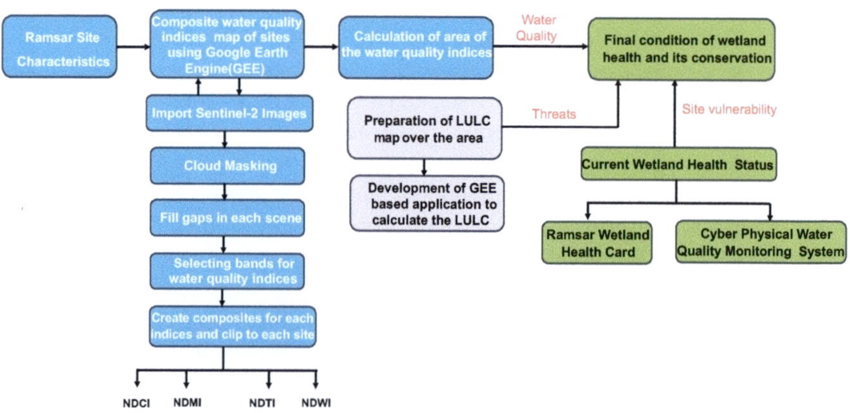

Fig. 2.2 Methodical schematic approach

2.3.2 Water Quality and Land Cover Assessment Methods

These were assessed using remote sensing techniques. Sentinel-2 images were processed through GEE to generate key remote sensing indices such as the NDCI and NDTI for assessment of eutrophicated and turbid mean areal extent between 2019 and 2024. These indices provide critical insights into various aspects of eutrophication and turbidity across the Ramsar sites. The data were collected annually, allowing for the creation of indices for the annual areal extent of maps to track changes over time. Apart from the maps, time series charts of the mentioned indices were calculated. The charts give a better idea about the conditions of the maps as they contain the monthly or sometimes weekly variation in the mentioned properties. The land cover for these sites was assessed using the ESA-World Cover with a ten-meter spatial resolution for the period 2020–2021 using the GEE [28]. It will be used to compute the built-up and cropland area across these sites in order to assess the threats of human activities in the wetland surroundings.

2.3.3 Criteria for Evaluating Water Quality in Different Wetland Types

Wetland water quality assessment is a specific process that involves physical, chemical, and biological factors. Among the most crucial parameters related to water quality are the level of nutrient concentration, suspended solids, the dissolved oxygen level, the pH, the turbidity, and the level of contaminants. Further, assessment of wetland health of these aquatic ecosystems is considered based on water quality, site management plan, wetland encroachment, biodiversity, etc., as shown in Table 2.2. The evaluation criteria for the wetland also had to depend on the type of wetland. For freshwater wetlands, nutrient levels and turbidity were given more focus owing to the great impact this has on aquatic plant and animal life. For saline or brackish wetlands, salinity and dissolved oxygen levels are affected as they adapt to the species adapted to these conditions. The information sheets of the Ramsar provided very useful data to classify and evaluate these types of wetlands by international standards [29, 30].

2.3.4 Statistical and Analytical Tools Used for Data Interpretation

The study is a multilateral approach to the complete understanding of wetlands. All Ramsar information sites were compiled together in detail to know about the peculiarities of the wetlands, such as ecological significance, biodiversity, hydrology, and socio-economic significance. All this data is a basis for further analysis. GEE was used to prepare, assess, and visualize the spatiotemporal assessment for the various

Table 2.2 Parameters considered for evaluating the wetland health [31]

Percentage of wetlands converted to non-wetland since 2000
Ratio of number of natural inlets choked and diverted to total number of natural inlets
Ratio of number of natural outlets choked and diverted to total number of natural outlets
Percentage of samples conforming to desired BOD/DO/COD levels
Percentage of wetland area covered by invasive macrophytes
Annual water bird count as a proportion of average count of last 5 years
Clearly demarcated wetlands map
Wetland management plan
Wetland notification

water quality parameters, such as eutrophication, turbidity, freshwater content, and some other important indices. Lastly, land use and land cover were assessed in the landscape context around it to determine human actions and impacts or pressures on the wetland. All these indices and land cover areal extent were assessed in the GEE. However, while preparing the spatial maps, we can use QGIS open-access software.

2.3.5 Limitations and Considerations in Data Collection and Analysis

This study provides valuable insights into wetland health and water quality. However, several factors cause limitations in the data collection and assessment. A major issue was the reliance on satellite images for continuous monitoring. Although they have advantages, they also come with limitations. One drawback is that the validity of satellite-derived indices can be compromised by atmospheric conditions, such as cloud cover or haze, which may distort reflectance values. Alternatives to address this limitation include using atmospheric correction algorithms, like the Sentinel-2 Level-2A products, or employing multi-sensor approaches that combine data from other satellites, such as Landsat or MODIS, which can provide complementary information. The availability of historical data and government reports could also be improved. While Ramsar information sheets contained extensive water quality and climatic data, insufficient records or conflicting information from various sources sometimes created problems. In such cases, integrating ground-truthing methods or citizen science initiatives could enhance data reliability and fill in gaps. Additionally, the variability in water quality characteristics and land use changes was primarily driven by anthropogenic inputs, such as agricultural runoff and urbanization, which are difficult to measure accurately. Implementing real-time monitoring using field-based sensors and cyber-physical systems can provide continuous data on water quality, allowing for a more comprehensive understanding of these impacts. Finally, temporal issues in data collection throughout the year, particularly regarding seasonality, posed additional challenges. Wetlands typically experience significant seasonal changes in water level, temperature, and species composition that an annual data collection schedule may not capture effectively. To address this, utilizing high-frequency monitoring with drones or unmanned aerial vehicles (UAVs) can provide more detailed temporal resolution, allowing for a better understanding of these dynamics. Therefore, conclusions drawn from this study must be contextualized within these natural temporal and spatial variations.

References

1. MoEFCC, *India's Wetland Wonder* (Government of India, 2024a). https://pib.gov.in/PressNote Details.aspx?NoteId=152029&ModuleId=3®=3&lang=1
2. N. Bassi, M.D. Kumar, A. Sharma, P. Pardha-Saradhi, Status of wetlands in India: a review of extent, ecosystem benefits, threats and management strategies. J. Hydrol. Reg. Stud. **2**, 1–19 (2014). https://doi.org/10.1016/j.ejrh.2014.07.001
3. MoEFCC, *COP13 National Reports* (2021b). https://www.ramsar.org/sites/default/files/doc uments/library/cop14nr_india_e.pdf
4. MoEFCC, *India's Wetlands of International Importance* (Government of India, 2024b). https:// indianwetlands.in/wetlands-overview/indias-wetlands-of-international-importance/
5. R.T. Kingsford, A. Basset, L. Jackson, Wetlands: conservation's poor cousins. Aquat. Conserv. Mar. Freshwat. Ecosyst. **26**(5), 892–916 (2016). https://doi.org/10.1002/aqc.2709
6. PIB, *Ministry of Environment, Forest, and Climate Change Submits Proposals for Wetland City Accreditation Under the Ramsar Convention on Wetlands for Cities of Indore, Bhopal and Udaipur* (MoEFCC, 2024). https://pib.gov.in/PressReleaseIframePage.aspx?PRID=199 3224#:~:text=Udaipur%3A
7. N. Kumar, U. Gupta, Y.V. Jhala, Q. Qureshi, A.G. Gosler, F. Sergio, GPS-telemetry unveils the regular high-elevation crossing of the Himalayas by a migratory raptor: implications for definition of a "Central Asian Flyway." Sci. Rep. **10**(1), 15988 (2020). https://doi.org/10.1038/ s41598-020-72970-z
8. K. Vaithianathan, *Flyways and Migratory Birds of India* (2022)
9. C.H. Besley, G.F. Birch, Deepwater ocean outfalls: a sustainable solution for sewage discharge for mega-coastal cities (Sydney, Australia): a synthesis. Mar. Pollut. Bull. **145**, 675–677 (2019). https://doi.org/10.1016/j.marpolbul.2019.06.010
10. U.K. Sarkar, B. Das Ghosh, M. Puthiyottil, A.K. Das, L. Lianthuamluaia, G. Karnatak, A. Acharya, B.K. Das, Spatio-temporal change analysis of three floodplain wetlands of eastern India in the context of climatic anomaly for sustainable fisheries management. Sustain. Water Resour. Manage. **7**(3), 41 (2021). https://doi.org/10.1007/s40899-021-00529-5
11. M.K. Goyal, S. Kumar, A. Gupta, *AI for Water Policy* (2024a), pp. 41–53. https://doi.org/10. 1007/978-3-031-72014-7_4
12. MoEFCC, *COP 14* (2021a). https://www.ramsar.org/sites/default/files/documents/library/cop 14nr_india_e.pdf
13. P. Singha, S. Pal, Wetland transformation and its impact on the livelihood of the fishing community in a flood plain river basin of India. Sci. Total Environ. **858**, 159547 (2023). https://doi. org/10.1016/j.scitotenv.2022.159547
14. R. Roy, M. Majumder, Assessment of water quality trends in Loktak Lake, Manipur, India. Environ. Earth Sci. **78**(13), 383 (2019). https://doi.org/10.1007/s12665-019-8383-0
15. A. Piralizefrehei, M. Kolahi, J. Fisher, Ecological-environmental challenges and restoration of aquatic ecosystems of the Middle-Eastern. Sci. Rep. **12**(1), 17229 (2022). https://doi.org/10. 1038/s41598-022-21465-0

16. J. Paonam, S. Chatterjee, Threat perception of stakeholders: case study of Loktak Lake, a Ramsar site under Montreux record in North East India. Wetlands **42**(8), 114 (2022). https://doi.org/10.1007/s13157-022-01630-x

17. UNESCO, *Sundarbans National Park* (World Heritage Site, 2024). https://whc.unesco.org/en/list/452/

18. M.K. Goyal, A.K. Gupta, S. Jha, S. Rakkasagi, V. Jain, Climate change impact on precipitation extremes over Indian cities: non-stationary analysis. Technol. Forecast. Soc. Chang. **180**(May), 121685 (2022). https://doi.org/10.1016/j.techfore.2022.121685

19. M. Garfí, A. Pedescoll, E. Bécares, M. Hijosa-Valsero, R. Sidrach-Cardona, J. García, Effect of climatic conditions, season and wastewater quality on contaminant removal efficiency of two experimental constructed wetlands in different regions of Spain. Sci. Total Environ. **437**, 61–67 (2012). https://doi.org/10.1016/j.scitotenv.2012.07.087

20. Y. Li, Y. Sun, J. Li, C. Gao, Socioeconomic drivers of urban heat island effect: empirical evidence from major Chinese cities. Sustain. Cities Soc. **63**, 102425 (2020). https://doi.org/10.1016/j.scs.2020.102425

21. V.L. Valenti, E.C. Carcelen, K. Lange, N.J. Russo, B. Chapman, Leveraging google earth engine user interface for semiautomated wetland classification in the Great Lakes Basin at 10 m with optical and radar geospatial datasets. IEEE J. Sel. Top. Appl. Earth Obs. Remote Sens. **13**, 6008–6018 (2020). https://doi.org/10.1109/JSTARS.2020.3023901

22. M.K. Goyal, C.S.P. Ojha, D.H. Burn, Nonparametric statistical downscaling of temperature, precipitation, and evaporation in a semiarid region in India. J. Hydrologic Engineering. **17**(5), 615–627 (2012)

23. A. Azhoni, M.K. Goyal, Diagnosing climate change impacts and identifying adaptation strategies by involving key stakeholder organisations and farmers in Sikkim, India: challenges and opportunities. Sci. Total. Environ. **626**, 468–477 (2018)

24. M.K. Goyal, S. Rakkasagi, S. Shaga, T.C. Zhang, R.Y. Surampalli, S. Dubey, Spatiotemporal-based automated inundation mapping of Ramsar wetlands using Google Earth Engine. Sci. Rep. **13**(1), 17324 (2023). https://doi.org/10.1038/s41598-023-43910-4

25. S. Ghosh, A. Das, Wetland conversion risk assessment of East Kolkata Wetland: a Ramsar site using random forest and support vector machine model. J. Clean. Prod. **275**, 123475 (2020). https://doi.org/10.1016/j.jclepro.2020.123475

26. E.R. Jones, M.F.P. Bierkens, N. Wanders, E.H. Sutanudjaja, L.P.H. Van Beek, M.T.H. Van Vliet, DynQual v1. 0: a high-resolution global surface water quality model. Geosci. Model Dev. **16**(15), 4481–4500 (2023)

27. M.K. Goyal, S. Kumar, A. Gupta, *AI for Water Treatment* (2024b), pp. 31–40. https://doi.org/10.1007/978-3-031-72014-7_3

28. D. Zanaga, R. Van De Kerchove, W. De Keersmaecker, N. Souverijns, C. Brockmann, R. Quast, J. Wevers, A. Grosu, A. Paccini, S. Vergnaud, O. Cartus, M. Santoro, S. Fritz, I. Georgieva, M. Lesiv, S. Carter, M. Herold, L. Li, N.-E. Tsendbazar, O. Arino, ESA World Cover 10 m 2020 v100. Zenodo (2021). https://doi.org/10.5281/zenodo.557193

29. V. Jain, K.S. Rautela, M.K. Goyal, Ecological restoration: an overview of science and policy regime. Ecosyst. Restor. Towards Sustain. Resilient Dev. 1–27 (2023)

30. Ramsar, *Site Management Plan* (Ramsar Convention, 2025). https://rsis.ramsar.org/ris-search/?language=en&f%5B0%5D=regionCountry_en_ss%3AIndia&f%5B1%5D=wetlandTypes_en_ss%3AMarineorcoastalwetlands&f%5B2%5D=managementPlanAvailable_i%3A-1&pagetab=0

31. Wetlands of India Portal, *Wetlands Health* (Wetlands of India Portal, 2024). https://indianwetlands.in/

Chapter 3
Coastal Ramsar Wetlands Water Quality Assessment

Abstract In this section, we present an overview of Coastal wetlands and their ten subtypes. Consequently, we present the ecological and economic significance of these sites in association with their role in mitigating the impacts of climate change. The study presents the regional biodiversity, water quality, land cover patterns, and wetland health for all sixteen coastal Ramsar sites. It is observed that Pallikaranai Marsh Reserve Forest has the lowest wetland health score of 44.44%. However, Vedanthangal Bird Sanctuary and Bhitarkanika Mangroves observed maximum eutrophicated and turbidity areas of 89.28 and 42.57%, respectively. The impact of human activities through pollution, land reclamation, and coastal development was assessed by comparing national site management with Ramsar and Global standards. Lastly, the study presents innovative technology integration to enhance the efficiency of coastal wetland site management.

Keywords Coastal wetlands · Water quality · Wetland health · Site management

3.1 Overview of Coastal Wetlands

Coastal wetlands are categorized as inundated and vegetated ecosystems that have distinctive hydrology and biogeochemistry impacted by seawater levels at different temporal periods of tidal waves over millennia. These vegetated ecosystems developed in the transition between marine and terrestrial ecosystems. They have endemic biodiversity that flourishes in the stressful environments of water-logged soils, saline waters, and anaerobic conditions. These wetlands can develop in various climatic conditions and distinctive eco-geomorphic regions like lagoons, estuaries, deltas, and oceanic islands. In the surroundings of these wetlands, mangroves grow along

---

The original version of the chapter has been revised. A correction to this chapter can be found at
https://doi.org/10.1007/978-3-031-96818-1_6

M. K. Goyal et al., *Monitoring India's Ramsar Wetlands*,
SpringerBriefs in Applied Sciences and Technology,
https://doi.org/10.1007/978-3-031-96818-1_3

the coastline, which provides resilience for hazards like cyclones, tsunamis, storm surges, etc. All the Indian Ramsar coastal wetlands have mangroves as the dominant vegetation in their surroundings. For example, the Sundarbans wetland has the maximum mangrove forest area in the world. These are considered essential components in estimating global greenhouse gas emissions with different climatic scenario modeling and assessing both continental and oceanic mass balances [1, 2].

The coastal wetland has been classified into ten subtypes as per the Ramsar Convention, which is shown in Table 3.1. As per the classification, the Intertidal forested, mud, sand flat, and marsh wetlands are regions surrounded by forest areas, exposed areas comprising mud and sand flats, and continuously drained by tides with the presence of marsh vegetation, respectively. The estuarine waters with nutrient-rich ecosystems comprise transition regions with freshwater from rivers mixed with the ocean. The saline lagoons are regions where coastal wetlands are partly separated from the sea through barriers and beaches and present parallel to the shoreline. Sandy and rocky shores are formed through loose and rocky material, respectively, and are present from the upper berm to the lower water level. Shallow marine waters are the regions where saline water accumulates at shallow depths near the coastline and evaporates, leaving the salt. The marine subtidal aquatic beds are the regions beneath the water surface that support the growth of submerged aquatic vegetation like seagrasses and algae. These wetlands have significant ecological and economic importance, as discussed in the subsequent section.

Table 3.1 Subtypes of coastal wetlands in India [3]

Coastal wetland subtypes	Numbers	Area (in ha)
Intertidal forested wetlands	7	268,930
Estuarine waters	5	650,191
Coastal brackish/saline lagoons	5	648,152
Sand, shingle, or pebble shores	4	224,801
Intertidal mud, sand, or salt flats	4	120,801
Permanent shallow marine waters	3	95,973
Coastal freshwater lagoons	3	206,900
Intertidal marshes	2	66,248
Marine subtidal aquatic beds (Underwater vegetation)	1	65,000
Rocky marine shores	1	4801

3.1.1 Ecological and Economic Importance of Coastal Wetlands

The coastal wetlands have multiple ecological and economic values that enhance the regional economy and conserve biodiversity. Ecologically, these sites are considered nurseries for various threatened marine species and hold invaluable habitats for birds, fishes, and invertebrates. They protect the shoreline from erosion and enhance the buffering impact of storms by absorbing excess floodwaters and having minimal impact on the coastal communities [4]. These sites have a high potential for carbon sequestration through seagrass beds, mangroves, and salt marshes in regional plants and soil with significant volumes of "Coastal Blue Carbon." Therefore, the preservation of these coastal wetlands prevents the release of this blue carbon into the atmosphere. It will regulate the regional climatic conditions and assist in mitigating the impact of climate change [5]. For example, the Thane Creek site stores 2688 tons of carbon annually, which has a value of Rs. 46 Lakh per annum as an ecosystem service apart from providing habitat for migratory birds like flamingos and other species [6]. Similarly, the estimated value of the blue carbon stored in the seagrass bed of the Gulf of Mannar site is Rs. 14 Lakh per annum [7]. The economic valuation comprising all the ecosystem services of the wetlands varies based on region, surrounding communities, and threatened biodiversity species [8]. All of these sixteen Ramsar sites have significant contributions to drinking water, groundwater recharge, tourism, and community development, specifically dependent on fishing activities. Chilika Lake is highly known for its fisheries, employing thousands of families. Sundarbans are an important resource mainly since they are a vital area for fisheries and honey production, and act as a natural cover against cyclonic storms. Eco-tourism is an emerging industry in the coastal wetlands of India, with bird watching and boat tours being seen as a source of attraction, especially to Vembanad-Kol and Chilika Lake sites.

3.1.2 Impacts of Climate Change

The changing climatic conditions exert substantial pressure on the coastal wetlands. These impact them through salinity intrusion with rising sea levels, changes in water quality, and altered hydrological conditions like changes in precipitation patterns with a rise in precipitation extremes [9, 10]. It degrades the regional biodiversity, mangrove vegetation, and impacts the livelihoods of the dependent fishing and agricultural communities. The rising sea level in the Sundarbans enhances the degradation of mangrove vegetation. Besides the surge in sea level, changes in regional temperature and precipitation patterns [11,12] are affecting the freshwater inflows into wetlands like Chilika Lake and Vembanad-Kol sites. It further enhances the pollutant concentration, which exacerbates eutrophication, which, on the whole, impacts aquatic life. The regional changes in water quality caused by climate change can shift the species composition within these wetlands, which in turn reduces their resilience concerning ecology and economic value. All these impacts of climate change and the rising anthropogenic influence across these sites required a transformation in the monitoring of the threats to Ramsar sites with the integration of remote sensing and artificial intelligence with a focus on wetland water quality and its overall health.

3.2 Water Quality Assessment

This section will present a discussion on the changing water quality patterns with an assessment of the sixteen Ramsar coastal wetland sites, as shown in Fig. 3.1, using remote sensing and artificial intelligence techniques. In this assessment, we initially reviewed the regional site characteristics and consequently discussed the mean areal extent of the eutrophicated and turbid area with the wetland health.

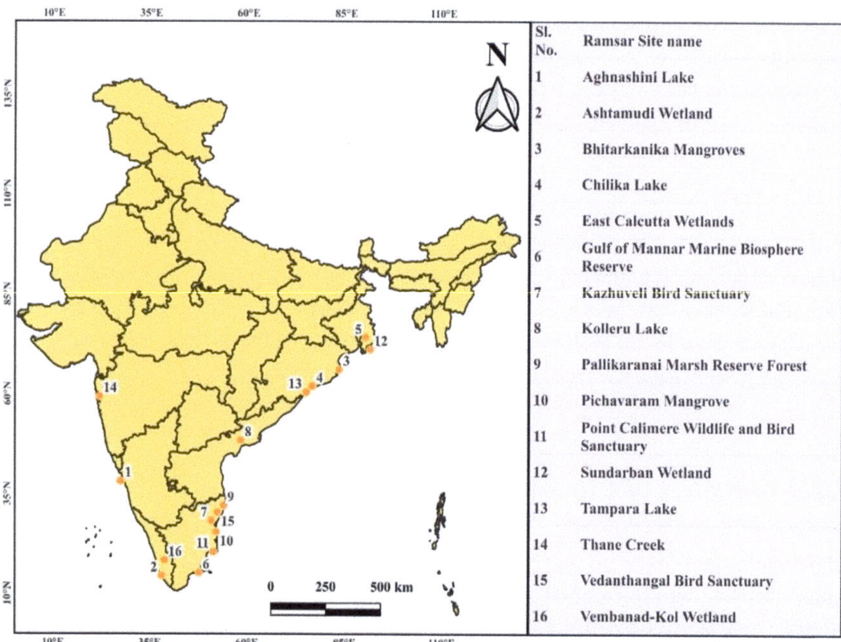

Fig. 3.1 Location map of coastal wetlands of India

3.2.1 Regional Water Quality Characteristics and the Land Cover Patterns

This section presents the details of each Coastal Ramsar site using Ramsar [3], along with land cover patterns and their water quality characteristics.

(a) **Aghnashini Wetland**

It was declared a Ramsar site in February 2023 as Aghanashini Estuary, Kumta Taluk, Uttara Kannada District, Karnataka, where the Aghanashini River meets the Arabian Sea. It stretches over an area of 48 km^2 and provides ecosystem services like controlling floods, erosion, and supports biodiversity in the brackish water ecosystem. The site has a tropical climate with saline water. It comprised a rich biodiversity of 84 fish species, five species of bivalves, 45 species of mangroves, and 117 species of birds, among which many are threatened taxa. Considering the Ramsar sites criteria, it was observed that the site was considered under the seven criteria among the nine designated Ramsar criteria. Overall, it was observed that the site is of great importance in supporting rare species, crucial habitats, and the spawning of fish, together with a rich biodiversity. It also supports more than 20,000 water birds and over 1% of the biogeographic population of nine bird species. A site management plan is supposed to achieve sustainable resource use, habitat restoration, and community involvement. Water quality should be monitored in the parameters of pH, salinity, and dissolved oxygen to achieve ecological balance. Based on the remote sensing assessment, it was observed that the site has a built-up and cropland area of 296.37 and 1673.01 km^2, respectively, in their surroundings. It shows a large extent of agricultural usage in the wetland surroundings. The wetland surroundings are highly populated, with several villages scattered all around the wetland, increasing possible anthropogenic pressures on the ecosystem. Overall, remote sensing and artificial intelligence assessment in the Google Earth Engine (GEE) showed that a site with 11.89 and 10.86% eutrophicated and turbid areas, and a 73.33% wetland health score were observed. Fishing activities pose substantial risks related to wetland health. Considering this, it was observed that water quality should be monitored regularly to establish eventual changes resulting from agricultural runoff, and appropriate regulations for fishing activities should be established and followed.

(b) **Ashtamudi Wetland**

This Ramsar site at Kollam, Kerala, is the state's second-largest estuarine system, spanning with an overall catchment area of 61.42 km^2. Its shape resembles a palm structure, with eight important arms merging into the Lakshadweep Sea. The wetland hosts 43 species of marshy and mangrove plants, with endangered species like *Zyzigium travencoricum* and three mangrove species. Additionally, it accommodates 57 species of birds and 97 species of fish, and thus, it is one of the most important spawning and passage grounds for many species of fish. The site is designated with four Ramsar criteria, like 1, 2, 3, and 8, as it holds hydrological functions, biological functions, and significant ecological functions. It was observed that the site has a built-up and cropland area of 24.35 and 37.32 km^2, respectively, in their surroundings. The site has 39.46 and 12.35% eutrophicated and turbid areas with a 93.33% wetland health score. The wetland suffers from severe pollution from the disposal of untreated sewage, contamination from the fish and seafood industries, and coconut husk retting. Sustainability in the estuary water qualities should, therefore, ensure the phasing out of sources of pollution. It includes sanitation improvement, industrial growth management, urban waste reduction, and protection of the ecological balance of the wetland. Seasonal changes and contaminants alter water quality.

(c) **Bhitarkanika Mangroves**

This site is spread over an area of 650 km^2 of the Brahmani and Baitarani Rivers deltaic region in the Kendrapara district of Odisha. This site ecosystem has a tropical climate with temperatures between 10 and 40 °C and an annual rainfall between 920 and 3000 mm. Bhitarkanika is the habitat of the highest density of Saltwater Crocodiles in India and hosts the largest nesting beach for the endangered Olive Ridley Sea Turtle. The sanctuary counts 174 species of birds and an annual nesting of over 20,000 birds; more importantly, it acts as a breeding ground for fish, mollusks, and other estuarine species. The mangrove vegetation has diverse species, like the *Avicennia marina* and *Ipomea pescaprae*. The site is considered under four Ramsar Criteria, i.e., 2, 4, 5, and 8, by supporting endangered species; it is a breeding location for birds and fish, and more than 30,000 waterbirds are reported regularly. It was observed that the site has a built-up and cropland area of 10.16 and 1067.64 km^2, respectively, in their surroundings. The site has 30.16 and 42.57% eutrophicated and turbid areas with an 82.22% wetland health score. Human activities like encroachment, pollution, poaching, and overfishing are significant threats to the wetland. Moreover, the threat of climate change looms large, as enhancement in sea levels and altered weather patterns can potentially have a disastrous impact on the mangrove ecosystems and their associated wildlife. An all-around conservation plan would need to consider human and environmental factors together.

The Odisha Forest Department manages them, but conservation strategies involve habitat restoration and community engagement that work toward maintaining the sustainability of the ecosystem as well as livelihoods.

(d) **Chilika Lake**

It is the biggest brackish water lake in India, spread over an area of 1165 km^2. It is located on the eastern coast of Odisha, predominantly in Puri District but partly stretching into Ganjam District; this giant lake stretches as far as 70 km in length with a maximum width of 30 km. Hydrologically, it connects with the Bay of Bengal and receives the Daya and Bhargavi Rivers. Depths range from 0.94 to 3.70 m. Chilka has a moderate climate, with temperatures varying from 53 to 95 °F and a mean annual precipitation of 888 mm. It has diverse fauna, such as the residents and migratory birds like ducks and flamingos, 158 species of fish and prawns, and marine life like dolphins. Among the flora species like *Halophyta ovata*, it is adapted to changing salinity levels. The site is qualified for Ramsar recognition on the basis of its environmental importance. It was observed that the site has a built-up and cropland area of 20.45 and 117.62 km^2, respectively, in their surroundings. The site has 17.47 and 23.80% eutrophicated and turbid areas with a 71.11% wetland health score. The lake is facing a threat of silt deposition, which has gradually been losing its depth and affecting its aquatic ecosystem. The inflow of silt entered the lake and finally settled their demands for immediate action through the concerned authorities. Further, human settlements around the lake should be inhibited to safeguard its ecological balance.

(e) **East Calcutta Wetlands**

This site, located near Kolkata city, is spread over an area of 125 km^2 with an average elevation of 2 m above sea level. This feature of using urban wastewater in sustainable fish farming, vegetable cultivation, and paddy fields instead of wastewater directly flowing into conventional treatment plants forms a signature feature of these wetlands. There are three significant wetland seasons: a cold season from mid-November until February with temperatures of almost 20.6 °C, a hot season from March to mid-June reaching temperatures of 40 °C, and a rainy season from mid-June until mid-September when the area receives heavy rainfall of 1200–1300 mm. The site has dominant species of plants, including *Eichhornia crassipes* and other *hydrophase* plants. The threatened fauna includes the following: marsh mongoose, small Indian mongoose, and palm civet. The site is considered under Ramsar Criterion 1 due to the innovative wastewater treatment system that supports sustainable agriculture and fisheries. It was observed that the site has a built-up and cropland area of 199.63 and 440.21 km^2, respectively, in their surroundings. The site has 39.12 and 25.69% eutrophicated and turbid areas with a 71.77% wetland health score. The site observed encroachment and development in wetlands due to the expansion of the urban areas, which.places a huge amount of pressure on them. Generally, the high rate of growth in Kolkata has placed greater stress on these wetlands and their natural resources, and disturbed traditional livelihood practices that maintained ecological balance. Thus, an integrated mitigation approach can be envisaged that better addresses the problems and improves conditions for the wetlands.

(f) **Gulf of Mannar Biosphere Reserve (GoMBR)**

The site is located between the districts of Toothukudi and Ramanathapuram, along India's southeastern coast. It occupies a total area of 526.72 km^2. The site is comprised of 21 islands, with corals forming the core zone of the Marine National Park. Seagrasses and mangroves surround the area, stabilize the coastline, filter pollutants, and prevent erosion. Seasonal oceanic currents due to the Southwest and Northeast monsoons influence the marine environment of the reserve. The site is the most biodiverse region in India, having 117 species of coral, over 450 species of fish, 160 species of birds, 641 crustaceans, and endangered species such as dugongs, whale sharks, green sea turtles, and dolphins. It is the only coastal site that qualifies for all the Ramsar criteria under the Ramsar Convention in India. It was observed that the site has a built-up and cropland area of 183.32 and 3626.91 km^2, respectively, in their surroundings. The site has 1.29 and 0.65% eutrophicated and turbid areas with an 80% wetland health score. There are critical threats from coastal development, pollution by agricultural runoff and domestic sewage, illegal fishing, and over-exploitation of marine resources. The site needs regulation of coastal development and initiation of habitat restoration programs to mitigate these threats and protect the region's biodiversity.

(g) **Kazhuveli Bird Sanctuary**

The site is located in the Tamil Nadu state of India in an area of 51.51 km^2 and was declared a Ramsar site in 2024. It is one of the largest wetlands in peninsular India, associated with the Bay of Bengal through the Uppukalli Creek and the Edayanthittu estuary. It falls along the Central Asian Flyway and hence serves as a crucial stopover for migratory birds and a breeding habitat for species of resident birds and fish. The site was considered a Ramsar site under the seven criteria of the Ramsar Convention. The sanctuary supports more than 750 species of flora and fauna, including 229 bird species, 85 fish species, and 72 butterfly species. It holds a few species classified as vulnerable or endangered, such as the Indian Flap-shelled Turtle and Indian Pangolin, which are critically necessary habitats for them. It was observed that the site has a built-up and cropland area of 3.06 and 68.45 km^2, respectively, in their surroundings.

The site has 42.37 and 28.79% eutrophicated and turbid areas with a 77.77% wetland health score. The site is now being subjected to multi-faceted threats such as salinization and aquaculture, urbanization, and agricultural and industrial pollution. The water quality exhibits seasonality due to the changes; during the dry months, the water possesses a higher concentration of salt, while during the monsoon periods, it has a lower concentration of salt. At present, habitat management and control of introduced species are used for conservation. A whole management plan is also under design.

(h) **Kolleru Lake**

The site is located between the Godavari and Krishna River basins in the Andhra Pradesh state and is spread over a 901 km^2 area. The site is about 55 km to the east of Vijayawada and acts as a flood-balancing reservoir. It is characterized by its geomorphology of lagoons, tidal marshes, and old beach ridges, and hence, the lake harbors varied terrestrial and aquatic ecosystems. The site has extensive hydrophyte vegetation and harbors floating species like Ipomea aquatica and Eichhornia crassipes, as well as submerged species like *Ottelia alismoides*. It is located in a semi-arid climate with an annual rainfall of 70–100 cm. The regional temperature ranges between 19 and 40 °C. Over 50,000 waterfowl make their homes in this lake, and migratory birds like *Open-Billed Storks* and *Grey Pelicans* visit here. The site is considered under the five Ramsar criteria, and it was observed that the site has a built-up and cropland area of 42.54 and 514.71 km^2, respectively, in their surroundings. The site has 58.53 and 3.50% eutrophicated and turbid areas with a 62.22% wetland health score. The intrusion of saline water during the dry season while allowing low freshwater inflow seriously hampers the lake's ecosystem. Agricultural runoff coupled with industrial pollution further endangers the lake. Efforts toward conservation, therefore, should be directed toward controlling sedimentation, improving the quality of water, and conserving biodiversity.

(i) **Pallikaranai Marsh Reserve Forest**

The site is located 20 km south of Chennai city in India, with a spread of 12.47 km^2. It is the only freshwater marsh in the region, aiding in flood management; it acts as a water buffer that lets in excess stormwater during wet spells and releases it back into the environment during dry spells, thus facilitating groundwater recharge. The marsh supports a heterogeneous ecosystem, including grasslands, shrublands, and water-filled depressions, which are important for hydrological stability in that region. The marsh is a biodiversity hotspot with a number of 381 species of flora and fauna, comprising 115 bird species, including the glossy ibis and gray-headed lapwings, besides being a habitat for significantly endangered species such as Russell's.viper. Among its flora lie 141 plant species and 29 species of grass. The site also provides critical habitats for several fish species, wherein it often serves as spawning grounds and supports their life cycles. The site is considered under the seven Ramsar criteria, with built-up and cropland areas of 74.35 and 17.38 km^2, respectively, in their surroundings. The site has.45.36 and 30.38% eutrophicated and turbid areas with a least 44.44% wetland health score. The untreated sewage and solid waste are major pollutants that have been identified as risks to the marsh, disturbing water quality and biodiversity. As evident in such areas and beyond, habitat loss and destruction due to urbanization and unauthorized development have occurred. Good settlement management in the surrounding areas with rigid pollution control is required to preserve the ecological integrity of the marsh.

(j) **Pichavaram Mangrove**

The site was declared a Ramsar site in April 2022, with an area of 14.78 km^2. It is located between the Vellar and Coleroon estuaries in Tamil Nadu state. Its mangrove ecosystem consists of 51 islets connected by creeks and channels of fresh as well as tidal water. The site hosts rich biodiversity, as well as a feeding and breeding ground for many migratory birds. Critically endangered species like the great, white-bellied heron and the spoon-billed sandpiper are found here, along with other vulnerable species such as the Olive Ridley turtle that nest annually along the Pichavaram coast. The area supports about 1,000 fishing families considered under the eight criteria of the Ramsar Convention. The site has built-up and cropland areas of 0.94 and 25.02 km^2, respectively, in their surroundings. The site has 50.62 and 16.31% eutrophicated and turbid areas with a 55.55% wetland health score. The main threats to the mangrove ecosystem include pollution, coastal development, climate change, and destructive fishing practices. The area focuses on conservation and should be put on habitat for mangroves, as well as the regulation of fishing to ensure the sustainability of the system.

(k) **Point Calimere Wildlife and Bird Sanctuary**

The site comprised the Point Calimere Forest, the Great Vedaranyam Swamp (GVS), and the Talaignayar Reserve Forest at the southern end of Nagappattinam District in

Tamil Nadu state. The site is at the head of the Cauvery Delta; the sanctuary has a monsoonal climate with annual rainfall between 1,000 and 1,500 mm, which is mostly contributed by the northeast monsoon, and temperatures vary between 25 and 34 °C. The site provides habitat for more than 30,000 flamingos, 200–300 endangered Grey Pelicans, the rare Spoonbill Sandpiper, and 119 waterbird species. It serves as an important breeding ground for marine fish species that are vital for coastal fisheries. The sanctuary hosts 257 birds, making it an important bird conservation area.

The site is considered under the four Ramsar criteria. Seasonal tidal actions and monsoons cause fluctuations in water salinity at specific areas and seasons, ranging from 5 to 50 ppt, thus enhancing the distinct ecological dynamics of the wetland. The site has built-up and cropland areas of 20.84 and 632.42 km^2, respectively, in their surroundings. The site has 15.94 and 39.10% eutrophicated and turbid areas with 84.44% wetland health score. The specific threats to the ecological character include agricultural runoff, industrial pollutants, overfishing, poaching, and unregulated grazing. Community involvement is a necessity in curbing overfishing and poaching, and proper handling of industrial waste is very much needed as a way to protect the environment.

(l) **Sunderban Wetland**

The site is part of the Ganges–Brahmaputra-Meghna River Delta; it covers some 42.60 km^2. with a wide spread of tidal waterways, mudflats, and small islands. It is located at the mouth of the Bay of Bengal; the region of the site stands relatively at a very low elevation of not more than six meters above sea level and has a tropical monsoonal climate with heavy rainfall during the monsoon season and dry weather for the rest of the year. Average annual rainfall is in the range of 1,600–1,800 mm. Surface water temperature lies between 24.63 and 27.89 °C. The pH value of the soil ranges from slightly acidic to alkaline and is recorded in the range of pH 5.4 and 8.5. The wetland is renowned for its biodiversity and is home to endangered species such as the Bengal tiger and the Ganges River dolphin, besides about 200 bird species, the estuarine crocodile, the Indian python, and approximately 250 species of fish. Flora consists of 34 true mangrove species and 40 associated plant species. Sundarban Wetland supports various life stages of such a diverse range of species, which makes it important as a protected area. The site is considered under the four Ramsar criteria. The site has built-up and cropland areas of 38.78 and 2949.71 km^2, respectively, in their surroundings. The site has 33.4 and 4.24% eutrophicated and turbid areas with a 91.11% wetland health score. The threats to the wetland include industrial discharge, agricultural runoff, encroachment, poaching, illegal logging, and conversion into agricultural areas. A comprehensive management plan for the site should consider local people in conservation to support sustainable activities and bolster anti-poaching efforts.

(m) **Tampara Lake**

The site is located near the Chatrapur town in Odisha's state, which lies within the Rushikulya River basin. It covers about 3 km^2 with a length of 5.8 km and a width of 6.7 m. It is parallel to the coastline, about one-kilometer distance from it. This lake was established from a.depression that occurred in the year 1766, and it has both coastal and Deccan Peninsula biogeographic region characteristics that promote rich biodiversity. It supports 60 species of birds, 46 species of fish, and 48 phytoplankton species. The important species include the threatened *Cyprinus carpio* (Amur carp) and *Sterna aurantia* (river tern). It also provides a habitat mainly for fish species, including *Labeo rohita* and *Cirrhinus mrigala*. The site is considered under the four Ramsar criteria. The site has built-up and cropland areas of 3.53 and 10.73 km^2, respectively, in their surroundings. The site has 26.02 and 4.73% eutrophicated and turbid areas with a 60% wetland health score. The wetland area is prone to pollution, overfishing, and habitat destruction as a result of rapid urbanization and intensification of agriculture. In terms of conservation strategies, measures toward controlling pollution and habitat restoration should be taken to protect the species and restore ecological balance.

(n) **Thane Creek**

The site is spread over the districts of Thane, Mumbai, and Mumbai Suburban in Maharashtra state and extends 12 km from the Ulhas River to the Vashi bridge. It has a fringe of mangroves and forms part of the humid bioclimatic zone, which gives it its formative importance in guarding the coast against cyclones and in checking the undue encroachment of seawater into the precious land. The creek is rich in diversity, and almost 20% of mangrove species exist here in India, along with 202 birds, 18 fish species, and scores of invertebrates, insects, and plankton. It is one of the few sites in the Central Asian Flyway for migratory birds and a site providing their essential nesting and foraging grounds to species such as *Acerodon celebensis*, *Oreochromis mossambicus*, *Sterna aurantia*, and others. It is observed that more than 244,000 waterbirds visit the site annually. The site is considered under the eight Ramsar criteria. The site has built-up and cropland areas of 70.63 and 17.21 km^2, respectively, in their surroundings. The site has 33.79 and 9.11% eutrophicated and turbid areas with a 68.88% wetland health score. Proximity to the urban centers of Mumbai, Thane, and Navi Mumbai regions exposes the creek to pressures of urban development, pollution, and other anthropogenic impacts.

(o) **Vedanthangal Bird Sanctuary**

The site is located in the Chengalpattu District of Tamil Nadu state, which has an area of 0.4 km^2 in the Palar-Cheyyar sub-basin. Its terrain is flat with clay soils and a 5-m-deep tank at the end of the bund to one side of the sanctuary. It supports an important site for ecological balance, mostly consisting of intertidal forest wetlands and having a tropical climate under the influence of the Northeast and Southwest monsoons. Rainfall is erratic, fluctuating between 400 and 1700 mm per year. The sanctuary has distinctive plant species such as *Alangium salviifolium*, *Albizia lebbeck*, and *Terminalia arjuna*. This site harbors more than 20,000 waterbirds, including the Threatened Black-headed Ibis, *Threskiornis melanocephalus*, and other species, such as the *Wood Sandpiper* (Tringa glareola) and *Eurasian Hoopoe* (Upupa epops). The site is considered under the seven Ramsar criteria. The site has built-up and cropland areas of 0.02 and 1.73 km^2, respectively, in their surroundings. The site has

89.28 and 5.82% eutrophicated and turbid areas with a 73.33% wetland health score. Ecological threats are significant across the site mainly due to human settlements and agriculture. Actually, the overall agricultural activities per annum crop cultivation, as well as perennial non-timber crops-worsen the situation. Thus, the growth of such crops must be restricted, and mineral content in the water and nutrient levels must be monitored so that eutrophic conditions are not promoted.

(p) **Vembanad-Kol Wetland**

The site has an area of 1512.50 km^2 on the southwest coast of India in Kerala state across the districts of Alappuzha, Ernakulam, and Thrissur. Its wetland system, which comprises the southern freshwater zone and the northern saltwater zone influenced by tidal cycles, is fed by the ten rivers from the Western Ghats. The site has tropical humid climate conditions that prevail in the wetland, with a mean yearly temperature of 24.2 °C. It also experiences major rainfall during the southwest monsoon season. It is also a biodiversity hotspot, hosting more than 141 bird species; among them, 50 are migrants and vulnerable species, such as the *Pelecanus philippensis*. It is also a breeding ground for waterfowl and supports rich aquatic life, especially breeding grounds for finfish, shellfish, and shrimps. The site is considered under the five Ramsar criteria. The site has built-up and cropland areas of 140.14 and 419.64 km^2, respectively, in their surroundings. The site has 37.08 and 21.68% eutrophicated and turbid areas with an 84.44% wetland health score. The site observed pollution due to industrial effluents, agrochemicals, and sewage, which caused it to suffer over-extraction of lime shells through lime shell fishery and excessive use of agrochemicals in agriculture. Since such effluents are also mixed before drainage into the lake, measures adopted should be effective enough to remove such pollutants before discharging water into the lake so that its ecosystem is properly sustained.

3.2.2 Comparative Assessment of Water Quality

Considering the water quality indices, the Normalized Difference Chlorophyll Index (NDCI) and Normalized Difference Turbidity Index (NDTI) were used to evaluate the levels of eutrophication and turbidity assessed using the Google Earth Engine (GEE). The eutrophication was primarily attributed to nutrient discharge from nearby croplands, while turbidity was linked to built-up areas, croplands, and sediment carried by runoff. The Vedanthangal Bird Sanctuary exhibited the highest percentage of eutrophicated areas (89.28%). Significant eutrophication was also observed at Kolleru Lake (58.53%) and Pichavaram Mangrove (50.62%), while the GoMBR had the lowest (1.29%). In terms of turbidity, the highest levels were recorded at Bhitarkanika Mangroves (42.57%), Point Calimere Wildlife and Bird Sanctuary (39.10%), and Pallikaranai Marsh Reserve Forest (30.38%). The GoMBR had the lowest turbidity (0.65%). Elevated turbidity, particularly at Bhitarkanika, often exceeds the limits set by the Central Pollution Control Board due to high sediment loads, particularly during the monsoon season. Both eutrophication and turbidity reduce photosynthesis and oxygen production, affecting the productivity, nutrient cycling, and food chains of wetland ecosystems.

3.2.3 Impact of Human Activities

The coastal Ramsar wetlands observed human activities like pollution, land reclamation, and coastal development activities. These activities lead to the degradation of habitat quality and threaten the biodiversity of these crucial ecosystems [13]. The descriptions of these activities are as follows:

1. **Pollution**

Ramsar sites are increasingly suffering from pollution stemming from various sources. Industrial effluents, which include chemicals, heavy metals, and untreated wastewater, are discharged into wetlands, degrading water quality and threatening aquatic ecosystems. Additionally, agricultural runoff from the overuse of fertilizers and pesticides contributes to nutrient pollution, leading to eutrophication. This process triggers algal blooms, depletes oxygen levels, and results in fish kills, adversely affecting species that depend on these habitats. Furthermore, urbanization around Ramsar sites has led to increased sewage disposal, exacerbating water contamination and disrupting the ecological balance of wetlands. Together, these factors pose significant threats to the health and sustainability of these vital ecosystems.

2. **Land Reclamation**

The process of land reclamation across the Ramsar sites causes Habitat loss as sites are frequently reclaimed for agricultural, residential, and infrastructural purposes, resulting in the loss of critical habitats for migratory birds, fish, and other wildlife. The hydrological Disruptions occurred due to reclamation activities that altered natural water flow patterns, impacting flood control, groundwater recharge, and the overall ecological functions of wetlands.

3. **Coastal Development**

The sites observe significant construction and infrastructure development. For example, mangroves and estuaries face degradation due to the development of ports, resorts, and residential areas, which reduces available habitats for various species. The increased tourism in Ramsar sites leads to habitat disturbances, pollution from littering, and expanded infrastructure, placing additional stress on fragile ecosystems.

Impacted Sites: The Vembanad-Kol Wetland is facing significant challenges due to industrial pollution and agricultural runoff, which are leading to water contamination and habitat degradation. Similarly, the Sundarbans are threatened by coastal development, including infrastructure projects and increased tourism, jeopardizing the mangroves and the rich biodiversity that includes the endangered Bengal tiger.

Chilika Lake is also impacted, as excessive fishing, land reclamation for aquaculture, and pollution disrupt the ecosystem, adversely affecting bird populations and fish diversity. Collectively, these human activities undermine the ecological integrity and sustainability of Ramsar sites, highlighting the urgent need for stricter regulations and improved management practices to ensure their long-term conservation.

3.2.4 Comparison with Ramsar Standards and Global Benchmarks

In order to assess the gaps in the present conservation plan for Indian Ramsar sites, we will compare them with Ramsar standards and global benchmarks. The comparison emphasizes the following important aspects:

1. **Conserving and Promoting Sustainable Use**

The Ramsar Convention emphasizes the principle of "wise use" of wetlands, advocating for a balanced approach that maintains both ecological health and human livelihoods while minimizing ecosystem damage through sustainable practices. Global benchmarks for wetland conservation, as observed in countries like the UK and Australia, demonstrate the effectiveness of integrative management strategies that seek to harmonize ecological integrity with human.activities such as eco-tourism and regulated land use [14]. In India, although several Ramsar sites exist, many are currently threatened by destructive activities, including overfishing, unregulated tourism, and pollution from industrial and agricultural runoff. Management plans for key sites, such as Vembanad-Kol and Chilika Lake, lag in the enforcement of these vital conservation principles, highlighting the urgent need for improved governance and sustainable management practices.

2. **Pollution Control**

Ramsar Standards emphasize the importance of addressing pollution through effective treatment of industrial effluents, efficient sewage management, and control over agricultural runoff. The goal is to implement preventive measures through robust regulatory frameworks, as promoted by Ramsar guidelines. In contrast, countries with extensive wetland ecosystems, such as the Netherlands and Japan, enforce much stricter regulations. Case studies from these nations illustrate the necessity of regulatory controls, which include advanced sewage treatment facilities, stringent restrictions on agricultural chemicals, and comprehensive monitoring programs to track pollutant movements. In India, however, Ramsar sites like Vembanad-Kol Wetland and Loktak Lake are severely impacted by pollution. Despite having a legal framework in place, ineffective enforcement and a lack of monitoring mechanisms allow pollution to persist, undermining the conservation efforts intended to protect these vital ecosystems [15].

3. Land Reclamation and Habitat Protection

Ramsar Standards emphasize that Ramsar sites should be safeguarded against destruction due to land reclamation and habitat loss, with guidelines underscoring the importance of preserving natural landscapes and ecosystems. In contrast, countries like New Zealand and Canada have implemented stringent restrictions on wetland reclamation, supported by large-scale restoration programs aimed at reversing historical losses of wetland areas. However, India faces significant challenges related to land reclamation, particularly in its extensive coastal Ramsar sites, such as the Sundarbans and Ashtamudi Lake. Here, urbanization, agriculture, and aquaculture threaten the integrity of wetland habitats. Although habitat restoration activities are underway, they are generally insufficient to counteract the rapid rate of habitat loss, highlighting an urgent need for more comprehensive and effective conservation efforts [16].

4. Coastal Development and Climate Change Resilience

The Ramsar Convention emphasizes the need to manage coastal development in a manner that minimizes irreversible changes, ensuring that wetlands continue to fulfill their crucial roles in coastal defense, flood control, and climate change mitigation. For example, in regions like Scandinavia, sustainable coastal development is effectively integrated with wetland protection to enhance resilience against rising sea levels and extreme weather events. These countries invest significant resources in improving their coastal wetlands to bolster their defenses against climate change. In contrast, Indian coastal Ramsar sites like the Sundarbans face substantial threats from human development and climate change. While management plans are in place, they often lack comprehensiveness regarding the scale of coastal development and its potential long-term impacts on the resilience of these vital wetlands. It underscores the need for more robust and inclusive strategies to protect these ecosystems.

3.3 Case Studies and Innovations in Coastal Wetland Management in India

Wetlands are critical ecosystems, providing sites for biodiversity maintenance, a buffer against flooding, and, most importantly, the cradle for water purification as well as contributing to climate regulation [17]. The scale of the damage done by pollution, land reclamation, and climate change at the frontline is, in many ways, enormous. Wetlands have always played and still play a crucial role in India's rich wetland heritage. In this section, we assessed the management of some of the coastal Ramsar sites in India. It outlines and addresses challenges, solutions, and the role that technology can play in the sustenance of these important ecosystems.

3.3.1 Case Studies

In this section, we discuss the case studies for selected Coastal Ramsar sites, which are as follows:

(a) **Vembanad-Kol Wetland—Trace Metal Contamination**

This site observed traces of metal contamination of elements like iron (Fe), zinc (Zn), cadmium (Cd), etc., in three zones: northern, central, and southern regions. The sediment samples reveal that metals like Cd, Zn, and Cu are more than the threshold levels. Ecological hazards result from these sediments. The metal concentration was the highest in the central zone because it is being affected by hospital waste and urban runoff. Seasonal analysis reveals that the metal concentration is higher during the monsoon season because of agricultural runoff and enhanced deposition from anthropogenic sources. The enrichment factor (EF) and geo-accumulation index (Igeo) demonstrated that severe contamination is present, especially for Zn and Cd, posing a hazard to organisms dwelling in sediment and could be influencing the food chain of the region. Specifically, in the Vazhakkal region, heavy metal Cd levels indicate significant pollution from fertilizers and pesticides. Heavy metals are generally cumulative in nature over the long term and thus impair the biogeochemical functions of wetlands. The study by [18] suggests how such a fragile ecosystem can be conserved from further degradation by the necessity of adopting sustainable management practices, such as controlling the discharge of wastes and reducing reliance on agrochemicals.

(b) **Ashtamudi Wetland: Introducing Willing to Pay (WTP)**

The study on Ashtamudi Lake in Kerala, India, by [19] presents an in-depth case study on the WTP of residents for wetland conservation. Ashtamudi Lake, being a site classified as a Ramsar site, provides several critical ecosystem services such as fisheries and aquaculture, water purification, and storm protection. Despite these benefits, significant degradation faces Ashtamudi due to pollution, mangrove deforestation, and unsustainable fishing practices. To estimate the feasibility of restoration, the authors surveyed the residents' WTPs for different restoration scenarios. The experiment is mainly concerned with three ecosystem services related to improvement in water quality, sustainable management of fisheries, and mangrove conservation. The restoration scenarios range from modest to ambitious in nature, each carrying different levels of improvement and a one-time payment. The results showed that mangrove protection was the most preferred while improvement in water quality and sustainably managed fisheries ranked the second and third preference, respectively. WTP toward minor restoration was found to be INR 5,259 per household, whereas those toward moderate and ambitious restorations stand at INR 6,489 and INR 8,503, respectively. It is interesting to note that the respondents who did not belong to the Kollam district but hailed from villages around the lake were relatively more willing to pay. The results suggest that even modest restoration efforts focused on water quality and mangrove conservation might generate sufficient economic benefits to outweigh the costs of implementation. It places restoration efforts not only as environmentally crucial but also economically viable for local communities. Last and most importantly, the study provides crucial insights into the preference of public views regarding environmental management and useful guidance for policymakers looking to build sustainable practices in Kerala's wetlands. This case study is an example of how economic valuation through WTP can influence conservation strategies and support sustainable development in ecologically sensitive areas.

3.3.2   Unique Challenges and Solutions in Water Quality Management

Water quality management of coastal wetlands in India faces unique challenges from various pollution sources, including industrial discharges, agricultural runoff, and unregulated aquaculture practices [20]. Additionally, coastal wetlands near industrial areas are at risk of contamination from heavy metals and chemical pollutants, which degrade water quality and harm aquatic life. Uncontrolled aquaculture further threatens ecosystems like Ashtamudi Lake, where untreated waste is released into the water, exacerbating pollution [21]. In order to address these challenges, several solutions can be implemented. Strengthening compliance with pollution control regulations is crucial; for instance, the gradual improvement in water quality at Ashtamudi Lake can be attributed to stricter regulations on aquaculture waste and industrial discharges. Restoration projects, such as mangrove and salt marsh rehabilitation at Point Calimere, have proven effective in trapping sediments and improving water quality [22], as these ecosystems naturally filter nutrients and reduce pollution. Moreover, community involvement is essential, as exemplified by initiatives in the Gulf of Mannar that encourage local participation in conservation programs [23]. These efforts promote sustainable exploitation of wetland resources and enforce stricter pollution control measures, ultimately contributing to the health of coastal wetlands.

3.3.3   Impacts on Ecosystem Health and Biodiversity

Pollution and habitat destruction of coastal wetlands pose significant threats to ecosystem health and biodiversity. Coastal Ramsar sites serve as crucial habitats for endangered species and migratory birds [24]. They are commercially important for fisheries, but habitat loss caused by poor water quality and land reclamation directly impacts these populations, leading to declines in biodiversity and ecosystem functionality. Migratory bird species, which rely on wetlands for breeding and feeding, are particularly vulnerable; for instance, altered hydrological conditions and saltwater intrusion have degraded habitats at Point Calimere, resulting in decreased bird population densities. Similarly, in the Gulf of Mannar, deteriorating water quality from coral mining and pollution has led to declines in marine biodiversity, fish populations, and coral cover [25]. Although some recovery has been observed due to conservation efforts, the overall situation remains precarious. Habitat fragmentation resulting from land reclamation for aquaculture further disconnects ecosystems, undermining the survival of species that depend on wetlands with diverse structures. Restoration and conservation activities, including reforestation and.pollution control is essential for enhancing ecosystem health and maintaining biodiversity in these vital coastal areas.

3.3.4 Contribution of Technology to Enhanced Management of Coastal Wetlands

Technology has significantly enhanced the management of India's coastal wetlands, introducing innovative solutions to longstanding challenges. Remote sensing and GIS play a crucial role in tracking changes in wetland health, allowing for the monitoring of water quality, land use, and habitat fragmentation. Satellite imagery, for instance, has been instrumental in assessing coral cover in the Gulf of Mannar and identifying areas impacted by pollution or overfishing. Additionally, hydrological modeling helps predict the effects of water management activities, such as the restoration of tidal flow, guiding sustainable management practices. Water quality monitoring technologies have also advanced. Wetlands like Ashtamudi Lake utilize real-time monitoring systems that employ automated sensors to track critical parameters such as dissolved oxygen, pH, and nutrient levels. This real-time data enables prompt responses to pollution events and enhances long-term water quality management. Bio-monitoring techniques, utilizing indicator species like clams and mussels, provide quantitative measures of ecosystem health and have been essential in tracking biodiversity restoration efforts in the Gulf of Mannar due to conservation initiatives [26]. Data-driven management strategies, including big data analytics and machine learning, are increasingly being adopted to process vast datasets from remote sensing and water quality sensors. These technologies can predict future trends in wetland health and inform adaptive management strategies [27]. Moreover, community-based data collection initiatives empower local populations to monitor wetland conditions through apps and citizen science, enhancing the accuracy and coverage of data, particularly in remote areas. Given the severe threats to India's coastal wetlands from pollution, land reclamation, and climate change, innovative management strategies that integrate community engagement, stringent pollution controls, and advanced technology have shown promise in mitigating these impacts in Ramsar sites like Point Calimere, Ashtamudi Lake, and the Gulf of Mannar. As India advances toward improved wetland management, leveraging these technologies and fostering sustainable livelihoods within communities will be vital for the long-term protection and sustenance of these critical ecosystems. Considering this, in the subsequent chapter, we discuss the Inland Ramsar wetlands water quality assessment of India.

References

1. C. Marchand, X. Ouyang, F. Wang, A. Leopold, Impact of climate change and related disturbances on CO_2 and CH_4 cycling in coastal wetlands. In *Carbon Mineralization in Coastal Wetlands* (Elsevier, 2022), pp. 197–231. https://doi.org/10.1016/B978-0-12-819220-7.00010-8
2. L. Windham-Myers, J.R. Holmquist, K.D. Kroeger, T.G. Troxler, Greenhouse gas balances in coastal ecosystems: current challenges in "blue carbon" estimation and significance to national greenhouse gas inventories, in *Balancing Greenhouse Gas Budgets* (Elsevier, 2022), pp. 403–425. https://doi.org/10.1016/B978-0-12-814952-2.00001-0
3. Ramsar, *Ramsar Sites Information Service* (Ramsar, 2024a). https://rsis.ramsar.org/ris-search/
4. S. Bhowmik, Ecological and economic importance of wetlands and their vulnerability: a review. Res. Anthol. Ecosyst. Conserv. Preserv. Biodivers. 11–27 (2022)
5. M.M. Moritsch, M. Young, P. Carnell, P.I. Macreadie, C. Lovelock, E. Nicholson, P.T. Raimondi, L.M. Wedding, D. Ierodiaconou, Estimating blue carbon sequestration under coastal management scenarios. Sci. Total Environ. **777**, 145962 (2021). https://doi.org/10.1016/j.scitotenv.2021.145962
6. B. Chatterjee, Hindustan Times. Hindustan Times (2020). https://www.hindustantimes.com/mumbai-news/carbon-sequestration-volume-from-thane-sanctuary-valued-at-46-lakh-per-year-maharashtra-government/story-pINilR6a9aLrQqI9q7XYKK.html
7. P. Kaladharan, P.U. Zacharia, S. Thirumalaiselvan, A. Anasukoya, L. Ratheesh, S.M. Sikkander Batcha, Blue carbon stock of the seagrass meadows of Gulf of Mannar and Palk Bay off Coromandel Coast, south India. Indian J. Fish. **67**(4) (2020). https://doi.org/10.21077/ijf.2020.67.4.98101-18
8. S. Singh, M.K. Goyal, E. Saikumar, Assessing climate vulnerability of Ramsar wetlands through CMIP6 projections. Water Resour. Manage **38**(4), 1381–1395 (2024). https://doi.org/10.1007/s11269-023-03726-3
9. A. Arabadzhyan, P. Figini, C. García, M.M. González, Y.E. Lam-González, C.J. León, Climate change, coastal tourism, and impact chains—a literature review. Curr. Issue Tour. **24**(16), 2233–2268 (2021)
10. E. Laino, G. Iglesias, Extreme climate change hazards and impacts on European coastal cities: a review. Renew. Sustain. Energy Rev. **184**, 113587 (2023). https://doi.org/10.1016/j.rser.2023.113587
11. S. Dubey, H. Gupta, M.K. Goyal, N. Joshi, Evaluation of precipitation datasets available on Google earth engine over India. Int. J. Climatol. **41**(10), 4844–4863 (2021)
12. M.K. Goyal, C.S.P. Ojha, D.H. Burn, Nonparametric statistical downscaling of temperature, precipitation, and evaporation in a semiarid region in India. J. Hydrol. Eng. **17**(5), 615–627 (2012)
13. M. Inácio, F.R. Barboza, M. Villoslada, The protection of coastal lagoons as a nature-based solution to mitigate coastal floods. Curr. Opin. Environ. Sci. Health **34**, 100491 (2023). https://doi.org/10.1016/j.coesh.2023.100491
14. B. Clarke, A.K. Thet, H. Sandhu, S. Dittmann, Integrating cultural ecosystem services valuation into coastal wetlands restoration: a case study from South Australia. Environ. Sci. Policy **116**, 220–229 (2021)
15. M.K. Goyal, S. Kumar, A. Gupta, *AI for Water Treatment* (2024a), pp. 31–40. https://doi.org/10.1007/978-3-031-72014-7_3
16. S. Rakkasagi, M.K. Goyal, S. Jha, Evaluating the future risk of coastal Ramsar wetlands in India to extreme rainfalls using fuzzy logic. J. Hydrol. **632**, 130869 (2024). https://doi.org/10.1016/j.jhydrol.2024.130869
17. V. Jain, K.S. Rautela, M.K. Goyal, Ecological restoration: an overview of science and policy regime. Ecosyst. Restor. Towards Sustain. Resilient Dev. 1–27 (2023)
18. A.R. Nisari, C.H. Sujatha, Assessment of trace metal contamination in the Kolwetland, a Ramsar site, Southwest coast of India. Reg. Stud. Marine Sci. **47**, 101953 (2021)

19. M. Sinclair, M.K. Vishnu Sagar, C. Knudsen, J. Sabu, A. Ghermandi, Economic appraisal of ecosystem services and restoration scenarios in a tropical coastal Ramsar wetland in India. Ecosyst. Serv. **47**, 101236 (2021). https://doi.org/10.1016/j.ecoser.2020.101236

20. M.K. Goyal, S. Kumar, A. Gupta, *Basics of AI for Water Management* (2024b), pp. 1–16. https://doi.org/10.1007/978-3-031-72014-7_1

21. M.A. Sreedevi, P.S. Harikumar, Occurrence, distribution, and ecological risk of heavy metals and persistent organic pollutants (OCPs, PCBs, and PAHs) in surface sediments of the Ashtamudi wetland, south-west coast of India. Reg.Nal Stud. Mar. Sci. **64**, 103044 (2023)

22. Dhan Foundation, *Assessment of Livelihood-Ecosystem Interdependencies for Integrated Management of "Point Calimere Ramsar Site"* Tamil Nadu, India. Madurai. (2021). Retrieved , from https://dhan.org/document/coastal-watch/Point Calimere Wetland Conserveation study report - CALL.pdf

23. K. Green Sea, M. Rajakumar, T. Umamaheswari, N.V. Sujath Kumar, P. Jawahar, N.R. Keer, R. Yadav, Ecosystem service approach for community-based management towards sustainable blue economy. Indian J. Anim. Sci. **91**(12), 1122–1126 (2021). https://doi.org/10.56093/ijans.v91i12.119843

24. A. Newton, J. Icely, S. Cristina, G.M.E. Perillo, R.E. Turner, D. Ashan, Anthropogenic, direct pressures on coastal wetlands. Front. Ecol. Evolution. Frontiers Media S.A. (2020, July 7). https://doi.org/10.3389/fevo.2020.00144

25. N.G.G. Asir, P.D. Kumar, A. Arasamuthu, G. Mathews, K.D. Raj, T.K.A. Kumar, Eroding islands of Gulf of Mannar, Southeast India: a consequence of long-term impact of coral mining and climate change. Nat. Hazards **103**(1), 103–119 (2020). https://doi.org/10.1007/s11069-020-03961-6

26. R. George, G.D. Martin, S.M. Nair, N. Chandramohanakumar, Biomonitoring of trace metal pollution using the bivalve molluscs, Villorita cyprinoides, from the Cochin backwaters. Environ. Monit. Assess. **185**(12), 10317–10331 (2013). https://doi.org/10.1007/s10661-013-3334-9

27. S. Pal, S. Debanshi, Machine learning models for wetland habitat vulnerability in mature Ganges delta. Environ. Sci. Pollut. Res. **28**(15), 19121–19146 (2021). https://doi.org/10.1007/s11356-020-11413-8

Chapter 4
Inland Wetlands Water Quality Assessment

Abstract In this section, we present an overview of Inland and Man-Made wetlands and their seventeen and eight subtypes, respectively. Consequently, we present the ecological and economic significance of these sites in association with their role in mitigating the impacts of anthropogenic activities. The study presents the regional biodiversity, water quality, land cover patterns, and wetland health for all sixty-nine inland Ramsar sites. It is observed that Deepor Beel has the lowest wetland health score of 37.80%. However, Pala Wetland and Keoladeo National Park observed maximum eutrophicated and turbidity areas of 98.39% and 91.48%, respectively. The impact of human activities through pollution, land reclamation, and site development was assessed by comparing national site management with Ramsar and Global standards. Lastly, the study presents innovative technology integration in order to enhance the efficiency of the Inland Wetland Site Management.

Keywords Inland wetlands · Water quality · Wetland health · Site management

4.1 Overview of Inland Wetlands

The Inland wetlands comprised marshes and wet meadows with a dominant flora of herbaceous plants, swamps, shrubs, and wooden swamps dominated by trees. These are present across the floodplains, isolated depressions surrounded by dry lands, along the margins of regional lakes, and low-lying regions where groundwater intercepts the soil surface or precipitation saturates the soil [1]. These wetlands are comprised of several man-made wetlands catering to the demands of the regional population. In India, there are several types of Inland and man-made wetlands, which are shown in Tables 4.1 and 4.2, respectively [2]. The natural inland wetlands are categorized primarily into permanent/intermittent with freshwater or brackish water lakes, marshes, streams, and peatlands dominated by shrub or forested vegetation. The lakes are considered substantial bodies of standing water present in distinct basins across natural depressions and fed by regional streams or rivers. The wetlands formed along the rivers lead to the creation of marshes and swamps, with a dominant

M. K. Goyal et al., *Monitoring India's Ramsar Wetlands*,
SpringerBriefs in Applied Sciences and Technology,
https://doi.org/10.1007/978-3-031-96818-1_4

vegetation of herbaceous plants and shrubs or trees, respectively. Sometimes, the rivers that are wider than the mapping unit are considered polygons or wetlands with significant importance in specific stretches, like the Upper Ganga River wetland between Brajghat and Garh Muketshwar region, and the Beas conservation reserve of the Beas River, with a 185 km stretch in Punjab, is considered a Ramsar site. While Man-made wetlands are primarily categorized as Reservoirs, canals, irrigated land, ponds, and salt sites. Reservoirs are considered ponds or lakes built for water storage by a dam site across a river, like Hirakud Reservoir in Odisha. Similarly, a Barrage was created for irrigation facilities along the Beas and Sutlej Rivers, leading to the formation of the Harike wetland in Punjab. The salt pans and wastewater treatment areas are prepared for the accumulation of saline or wastewater in order to prepare salt or treat wastewater using salt pans or constructed wetlands, respectively. Similarly, the aquaculture ponds are prepared for the breeding of freshwater or marine fisheries [2]. In addition to this, in certain conditions, the wetlands are formed with canal or agriculture water-logged regions due to unlined canals and excessive irrigation.

Table 4.1 Sub-types of inland wetlands in India [2]

Subtype	Numbers	Area
O: permanent freshwater lakes	19	172,205
Ts: seasonal/intermittent freshwater marshes/pools	15	158,265
M: permanent rivers/streams/creeks	12	232,776
Tp: permanent freshwater marshes/pools	10	80,907
P: seasonal/intermittent freshwater lakes	8	84,294
N: seasonal/intermittent/irregular rivers/streams/creeks	5	114,777
Q: permanent saline/brackish/alkaline lakes	5	58,004
Xf: freshwater, tree-dominated wetlands	5	29,932
W: shrub-dominated wetlands	4	4064
R: seasonal/intermittent saline/brackish/alkaline lakes	2	36,000
U: permanent non-forested peatlands	2	1395
Xp: permanent forested peatlands	2	1395
Y: permanent freshwater springs; oases	2	1395
Zk(b): karst and other subterranean hydrological systems	2	1395
Sp: permanent saline/brackish/alkaline marshes/pools	1	49
Ss: seasonal/intermittent saline/brackish/alkaline marshes	1	1248
Vt: tundra wetlands	1	49

Table 4.2 Sub-types of man-made wetlands in India [2]

Subtype	Numbers	Area
Water storage areas/reservoirs	32	185,860
Ponds	5	5020
Irrigated land	4	72,720
Aquaculture ponds	3	102,944
Seasonally flooded agricultural land	3	27,564
Salt exploitation sites	2	62,500
Wastewater treatment areas	2	13,130
Canals and drainage channels, or ditches	2	28,027

4.1.1 Ecological and Economic Importance of Inland Wetlands

Several Inland wetlands in the nation are seasonal and present across the arid and semi-arid regions. The amount of water present in them during the drier or wet period provides distinctive functions for the critical habitat of wildlife and breeding conditions on these sites [1, 3]. The economic values of these wetlands are significant in the context of biodiversity conservation, irrigation facilities, recharging groundwater, etc. However, their importance is significantly higher in the context of urban wetlands. It enhances the regional air quality, replenishes drinking water, minimizes runoff, and enables community development through recreation and tourism activities while enhancing their living conditions [4]. For example, Bhoj wetland, an urban wetland providing water supply for Bhopal city, has an economic valuation of Rs. 177.3 million per annum [5].

4.1.2 Impact of Climate Change and Human Activities

Climate change with rising warming conditions [6, 7] enhances the risk of wetland loss in future climatic scenarios, specifically in Northern India [8]. In the previous ten years, India has lost thirty-eight percent of its inland wetlands primarily due to heavy contamination with heavy metals and pesticides. The seasonal nature of various inland wetlands makes them dry in the summer season, and the encroachment, along with the dumping of waste in landfills, has prominently occurred in the Inland wetlands. The wetlands in northeastern states like Nagaland, Meghalaya, and Manipur observed the maximum content of organochlorine pesticides. For example, the high content of heavy metals and pesticides, more than 0.558 ppm, in Loktak Lake of Manipur state. It causes a decline in fish yield by more than 50%.

Lead and Zinc are the prominent heavy metals present in the wetlands of Uttar Pradesh, Bihar, and Haryana states [9].

4.2 Water Quality Assessment

This section will present a discussion on the changing water quality patterns with an assessment of the sixty-nine Inland Ramsar Wetland sites, as shown in Fig. 4.1, using remote sensing and artificial intelligence techniques. In this assessment, we initially reviewed the regional site characteristics and consequently discussed the mean areal extent of the eutrophicated and turbid area with the wetland health.

4.2.1 Regional Water Quality Characteristics and the Land Cover Patterns

This section presents the details of each Inland Ramsar site using Ramsar [2], along with their assessment for land cover patterns, water quality characteristics, and wetland health score.

(1) Ankasamudra Bird Conservation Reserve

This site is located in Karnataka state in an area of 0.99 km^2. It has a historic irrigation tank with thousands of gum Arabic trees, which is a vital nesting site for waterbirds. The reserve lies approximately 40 km from Hampi on the semi-arid Deccan Plateau and experiences hot summers, with milder and moderate rainfall in winter. It has varied wildlife populations, such as 240 bird species, 210 plant species, 41 fish

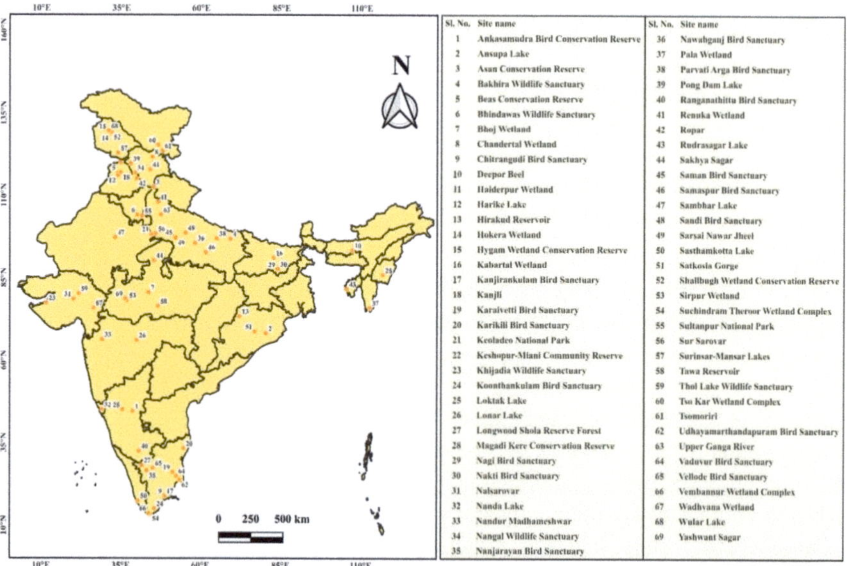

Fig. 4.1 Location map of inland Ramsar wetlands of India

species, 25 reptiles, and 81 medicinal plants. The site fulfills the Ramsar Criteria 2, 3, 4, 5, and 6. The site has a mean eutrophicated and turbid area of 63.49% and 16.81%, respectively. It also faces threats like habitat degradation, pollution, invasive species, and water management. The built-up and cropland land cover areas are 0.02 km^2 and 2.09 km^2, respectively, with a wetland health score of 66.7%. The management programs should be focused on restoration, biodiversity monitoring, and community involvement.

(2) Ansupa Lake

The site is Odisha's largest freshwater lake, covering 2.31 km^2, an oxbow lake created by the Mahanadi River in Cuttack district. It forms a small backwater connected to the Mahanadi through Kabula Nala and gets drained by Huluhula Nala. Rich biodiversity is seen here, hosting 194 species of birds and 61 species of fish, with several threatened species, including *Rynchops albicollis* (Endangered) and *Clarias magur* (Endangered). It fulfilled the Ramsar criteria 2, 7, and 8 since the lake has ecological importance. The site has a mean eutrophicated and turbid area of 47.99% and 4.85%, respectively. The site's surroundings have a built-up and cropland area of 0.03 and 6.22 km^2, respectively, with an overall wetland health score of 66.7%. The threats to the lake includes pollution, overfishing, and siltation. Management of habitats, eco-tourism, and sustainable fisheries are among the measures considered in the management of the lake.

(3) Asan Conservation Reserve

The site is located at the confluence of the rivers Asan and Yamuna in Dehradun, Uttarakhand. The site spreads across 4.44 km^2 and has varied riverine and lacustrine habitats, supporting more than 330 species of birds, such as the Ruddy Shelduck and Common Teal, and hosts a habitat for 49 species of fish, like commercially important Rohu and Catla. Thus, it satisfies five criteria (2, 3, 4, 6, 8) for site designation as a Ramsar site. The site has a mean eutrophicated and turbid area of 27.60% and 46.99%, respectively. The site's surroundings have a built-up and cropland area of 13.09 and 1.35 km^2, respectively, with an overall wetland health score of 100%. The site has the least vulnerability; however, threats like agricultural runoff, habitat loss, overfishing, and invasive species persist in the region. Proper habitat restoration, pollution control, sustainable fisheries, and community engagement are the emphasis of the management of this wetland.

(4) Bakhira Wildlife Sanctuary

The site is located in Uttar Pradesh state, spread across 28.94 km^2, and serves as a crucial wetland ecosystem. It supports around 80 waterbird species, including the vulnerable Sarus crane, and fish species such as Wallago attu. The sanctuary meets multiple Ramsar criteria, i.e., 2, 3, 4, 8, for its biodiversity, particularly in supporting vulnerable species and sustaining fish vital for local livelihoods. The site has a mean eutrophicated and turbid area of 41.42% and 13.19%, respectively. The site's surroundings have a built-up and cropland area of 2.11 and 63.68 km^2, respectively, with an overall wetland health score of 71.10%. The key threats observed include agricultural runoff, habitat encroachment, and invasive species like water hyacinths. The proper site management should focus on habitat restoration, water quality monitoring, sustainable fishing, and community engagement.

(5) Beas Conservation Reserve

The site stretched over 64.28 km^2 along a 185 km length of the River Beas in India. The site covers numerous districts in the state of Punjab. The habitats vary from islands and sandbars to wide pools in the riverine areas. It hosts a very diverse flora and fauna, including endangered species like the *Indus River Dolphin* and the critically endangered *Gharial*, thus fulfilling Criterion 2 for Ramsar designation. The river harbors over 90 species of fish, thus qualifying to meet Criterion 7 regarding large numbers of fish species. The site has a mean eutrophicated and turbid areas of 4.10% and 60.93%, respectively. The site's surroundings have a built-up and cropland area of 1,497.29 and 20,050.33 km^2, respectively, with an overall wetland health score of 93.3%. The site observes threats like pollution and illegal fishing, which cause habitat disturbance. The conservation approaches should involve monitoring biodiversity, repairing habitats, anti-poaching measures, and community involvement.

(6) Bhindawas Wildlife Sanctuary

The site spread over 4.12 km^2 in Haryana state, and it was declared a Ramsar Site in May 2021. With more than 265 bird species, the site supports some endangered species like the Egyptian vulture and the steppe eagle (Criterion 2), helps preserve regional biodiversity (Criterion 3), is a significant stopover for migratory birds (Criterion 4), and supports more than 1% of some waterbird populations (Criterion 6). The site has a mean eutrophicated and turbid area of 38.32% and 12.92%, respectively. The site's surroundings have a built-up and cropland area of 0.42 and 8.15 km^2, respectively, with an overall wetland health score of 66.7%. The site management should focus on dealing with agricultural pollution and clearing encroachments.

(7) Bhoj Wetland

The site is located in Bhopal, Madhya Pradesh state, and is composed of two artificial lakes: the Upper Lake (30.72 km^2) and the Lower Lake (1.29 km^2), with a total area of 32.01 km^2. It features a varied aquatic environment in a dry climate with temperatures as high as 10.4–44 °C and an average yearly rainfall of 1,179 mm. It is home to 106 macrophyte species, 43 fish species, and more than 160 bird species. The site meets Ramsar Criteria 1, 2, 3, 5, 8. The site has a mean eutrophicated and turbid area of 23.19% and 7.25%, respectively. The site's surroundings have a built-up and cropland area of 40 and 101.77 km^2, respectively, with an overall wetland health score of 77.8%. Site management should focus on keeping a check on polluting effluents from the surrounding population.

(8) Chandertal Wetland

This site is a high-altitude natural lake in the state of Himachal Pradesh's Lahaul and Spiti District at an elevation of 4,337 m, covering 0.49 km^2. It lies in a cold, arid region where human interference is minimal, with negligible built-up and cropland area in its surroundings, respectively. It retains pristine water quality, being primarily fed by glacial melt, hence mesotrophic conditions. The wetland faces extreme temperatures during winter (-37 to $-40\ °C$) and minimal rainfall due to its rain-shadow position. The site supports diverse flora and fauna, which also includes the endangered species, such as the Snow Leopard, makes the site fulfill Ramsar Criteria 2 and 3. The site has mean eutrophicated and turbid areas of 7.98% and 62.74%, respectively, with an overall wetland health score of 97.8%. The site observes the impacts of climate change and tourism as a potential threat, with proper regulation of tourism activities.

(9) Chitrangudi Bird Sanctuary

The site, also known as "Chitrangudi Kanmoli," is located in the Ramanathapuram district, Tamil Nadu, in an area of $2.60\ km^2$. The sanctuary holds significance for its winter migratory birds and local biodiversity. It receives water from rainfall, groundwater, and river runoff. Intermittent water levels occur due to drought. The wetland hosts about 50 species of birds, along with the spot-billed pelican and Asian openbill (Criterion 4), and hosts high populations of species like *Pelecanus philippensis* (Criterion 6). Regional biological diversity is also still intact in the site (Criterion 3). The site has a mean eutrophicated and turbid area of 57.36% and 38.71%, respectively. The site's surroundings have a built-up and cropland area of 0.13 and 3.81 km^2, respectively, with an overall wetland health score of 71.1%. The major threats to the site are groundwater extraction for agriculture and floodwater management, which require effective conservation planning.

(10) Deepor Beel

The site is spread over $40\ km^2$ as a freshwater lake located near Guwahati, Assam, serving as an essential stormwater basin and representing the Burma Monsoon Forest biogeographic province. It lies at an elevation of 53 m within a U-shaped valley formed by ancient geological processes. The wetland features diverse habitats, including deep open water, marshes, mudflats, and scattered forests, supporting a rich array of biodiversity. Its flora includes aquatic plants like water hyacinth and giant water lily, while the fauna comprises 50 fish species and globally threatened birds such as the Spot-billed Pelican, meeting Ramsar Criteria 1, 2, 4, 7, and 8. The site has a mean eutrophicated and turbid area of 20.58% and 2%, respectively. The site's surroundings have a built-up and cropland area of 0.33 and 8.31 km^2, respectively, with an overall wetland health score of 37.8%. The site faces key threats, including pollution, habitat degradation, and changes in land use. Since the wetland has a significant vulnerability due to low health card scores, it should be taken care of properly, eliminating all the threats.

(11) Haiderpur Wetland

The wetland spreads over an area of $69.08\ km^2$ and is in the Muzaffarnagar district of Uttar Pradesh state. The site was formed after the construction of the Madhya Ganga Barrage in 1984. The present wetland can be said to comprise habitats like deep reservoirs, shallow flooded areas, and river stretches. The humid subtropical climate prevailing over there influences its seasonal hydrology and biodiversity. It supports more than 200 species of birds, including the endangered Black-bellied Tern and Sarus Crane, and supports aquatic life, such as *Labeo rohita*. The site also meets Ramsar Criteria 2, 3, 4, 5, and 6, owing to its ecological importance, rarity, and high populations of waterbirds. The site has a mean eutrophicated and turbid area of 26.19% and 42.96%, respectively. The site's surroundings have a built-up and cropland area of 1.80 and 77.02 km^2, respectively, with an overall wetland health score of 68.9%. The present threats sites are facing include pollution, habitat loss, overfishing, and invasive species. Thus, management practices should involve controlling pollution and restoring sites.

(12) Harike Lake

The site spread over 41 km^2, is located at 31° 13′ N and 75° 12′ E in Punjab's Kapurthala, Ferozepur, and Amritsar districts. It was formed by damming the Beas-Sutlej watershed, which is Punjab's largest wetland with a designated wildlife sanctuary under the Indian National Wetland Programme. The lake supports over 196 bird species, including migratory and resident birds. It serves as a vital water source for irrigation and drinking. It meets Criteria 1 for its unique wetland characteristics and biodiversity. The site has a mean eutrophicated and turbid area of 16.85% and 54.47%, respectively. The site's surroundings have a built-up and cropland area of 6.37 and 123.67 km^2, respectively, with an overall wetland health score of 71.1%. The site management practices to be considered include removing invasive species and controlling pollution.

(13) Hirakund Reservoir

The site is situated between Laxmidungri and Chandili Dunguri hills in the Sambalpur district of Odisha state. The site spreads over 654 km^2 and is considered India's largest earthen dam, established in 1957. Its facilities include regional flood control and irrigation over 75,000 km^2. The water level in the reservoir changes with the season, the highest recorded to date being during the monsoons. It has a range of habitats that harbor over 130 species of birds and over 54 species of fish, including many endemics and species of economic importance besides sloth bears and leopards. The site meets the Ramsar Criteria 1, 2, 3, 4, 5, 6, and 8 for its exceptional wetland type, biodiversity, and support for migratory birds. The site has a mean eutrophicated and turbid areas of 5.70% and 9.16%, respectively. The site's surroundings have a built-up and cropland area of 88.38 and 1,902.80 km^2, respectively, with an overall wetland health score of 80%. The major threats the site is facing are pollution, overfishing, and unregulated tourism, which need proper management.

(14) Hokera Wetland

The site is in the Kashmir Valley, spreads over 13.75 km^2, and is a very important support system for the migratory routes of birds from Siberia, China, Central Asia, and Europe. It houses 68 bird species, including the endangered white-eyed pochard; it is an important spawning area for fish. Wetland Ramsar Criteria: The wetland meets criteria 2, 4, 5, and 8. It hosts endangered species and supports essential feeding and breeding habitats for migratory birds, as well as a spawning habitat. The site has a mean eutrophicated and turbid area of 16.44% and 44.48%, respectively. The site's surroundings have a built-up and cropland area of 11.41 and 45.63 km^2, respectively, with an overall wetland health score of 68.9%. The major threats the site faces in this regard are sedimentation, nutrient loading, and fluctuating water levels. The proper management efforts are aimed at habitat restoration, control of invasive species, and bird population monitoring to preserve the wetland's biodiversity.

(15) Hygam Wetland

The site is spread over 8.02 km^2 in Baramulla district, Jammu and Kashmir, and lies within the basin of the Jhelum River near Wular Lake. The hydrological services from this wetland include flood control and recharging aquifers, but it faces the threat of siltation and nutrient accumulation by willow plantations. The wetland supports a variety of bird species, such as Cyprinus carpio and Aythya ferina, and has received more than 20,000 waterbirds between the years 2015 and 2021. It fulfills the Ramsar criteria 1, 2, 3, 5, and 6. The site has a mean eutrophicated and turbid area of 26.71% and 41.63%, respectively. The site's surroundings have a built-up and cropland area of 1.15 and 13.34 km^2, respectively, with an overall wetland health score of 57.8%. The key threats the site faces include habitat degradation and the conversion of marshland to agriculture. Thus, conserving sites requires restoring their habitat and working with the people living around them.

(16) Kabartal Wetland

The site is also known as Kanwar Jheel, and it covers an area of 26.2 km^2 in the Begusarai district in Bihar state. This fresh marsh supports over 200 species of birds, like the endangered Greater Spotted Eagle and Baer's Pochard. The wetland qualifies under Ramsar Criteria 1, 2, 3, 4, 7, and 8 owing to its biodiversity and its supporting role to vulnerable species as well as fish spawning. The site has a mean eutrophicated and turbid area of 27.40% and 6.45%, respectively. The site's surroundings have a built-up and cropland area of 5.13 and 93.59 km^2, respectively, with an overall wetland health score of 82.2%. The main threatening processes of the site are agricultural expansion, abstraction of water, pollution, and invasive species. Thus, management initiatives for habitat restoration, pollution reduction, sustainable agriculture, and community participation need to be implemented.

(17) Kanjirankulam Bird Sanctuary

The site is in Tamil Nadu state and is spread over 1.04 km^2. This site can support a wide variety of flora and fauna, in addition to species such as the Barbet and Kingfisher. The identified Ramsar Criteria 1, 2, and 4 are met because of the important ecological functions, including recharging groundwater, flood control, and a living place for various species. It receives annual rainfall of 1,500–2,000 mm, sub-tropical forests, and wetlands. The site has a mean eutrophicated and turbid area of 50.54% and 41.36%, respectively. The site's surroundings have a built-up and cropland area of 0.02 and 0.56 km^2, respectively, with an overall wetland health score of 73.3%. The major hazards that sites face include siltation from the uplands and human impact.

The core of conservation programs focuses on control measures for soil erosion, check dams, and community education.

(18) Kanjli Wetland

The site is spread over 1.83 km^2 in the state of Punjab, India. It hosts varied biodiversity, thereby offering the required ecological services, including water recharging, quality improvement, and flood control. It was formed by the Kali Bein rivulet and houses flora like Phragmites and Typha, besides numerous fish and bird species. The site supports food chains and thus fulfills Ramsar Criterion 3. The site has a mean eutrophicated and turbid areas of 9.37% and 60.14%, respectively. The site's surroundings have a built-up and cropland area of 11.87 and 138.83 km^2, respectively, with an overall wetland health score of 73.3%. The major threats, site observing invasive species in the form of water hyacinths, have posed a significant threat along with deforestation and organic accumulation. Management efforts are mainly focused on the removal of invasive species, stabilization of the groundwater table, and improvement in water quality.

(19) Karaivetti Bird Sanctuary

The site is spread over an area of 4.53 km^2 and is in Tamil Nadu's Ariyalur District. It is located between 10° 58′ N and 79° 02′ E. The Forest and Public Works Departments manage the site. It receives water from the Mettur Dam and the northeast monsoon, which are vital in groundwater recharge and supporting agriculture. The sanctuary serves as an important stopover for migratory birds like Bar-headed Geese and Pintail Ducks, hosting 198 bird species. It meets Ramsar Criteria 2, 3, 4, 5, and 6 for supporting vulnerable species and over 20,000 waterbirds. The site has a mean eutrophicated and turbid area of 41.58% and 18.08%, respectively. The site's surroundings have a built-up and cropland area of 0.12 and 7.35 km^2, respectively, with an overall wetland health score of 77.8%. The key threats include habitat degradation, pollution, and water scarcity, with ongoing management efforts focused on protecting the wetland's biodiversity and water quality.

(20) Karikili Bird Sanctuary

The site lies in Tamil Nadu state and is located in an area of 0.58 km^2. It is located near the village of Karikili, about 8 km north of Vedanthangal. It is comprised of two non-perennial irrigation tanks at a level of roughly 100 m above sea level. The sanctuary satisfies Ramsar Criteria 3 and 4 because it is a diverse ecosystem with important breeding and feeding grounds, respectively. It also meets criteria 6 and 7. The site has a mean eutrophicated and turbid area of 79.07% and 15.41%, respectively. The site's surrounding has a built-up and cropland area of 0.01 and 0.61 km^2, respectively, with an overall wetland health score of 68.9%. The site suffers from conflict in its land use, and the management tasks accordingly focus on biodiversity conservation, supported by local community involvement and sustainable water management.

(21) Keoladeo Bird Sanctuary

The site is in Bharatpur, Rajasthan, covering 29 km^2, situated in the Indo-Gangetic Plains, it sported a monsoon-fed freshwater swamp that covered about 10 km^2 for most of the year, split up into units controlled by sluice gates. The extreme temperature fluctuations here range from 51 °C in summer to 0 °C in winter, while the mean annual rainfall is approximately 662 mm. Dry deciduous as well as aquatic plants and 353 birds, along with the sambar and nilgai, thrive here. It is recognized under criterion 2 only. The site has a mean eutrophicated and turbid area of 8.78% and 91.48%, respectively. The site's surroundings have a built-up and cropland area of 19.37 and 92.74 km^2, respectively, with an overall wetland health score of 86.7%. The site is being declared a national park, facing the most serious threat of poaching, grazing, and water scarcity, but it remains of great importance for the survival of birds.

(22) Keshopur-Miani Community Reserve

The site, located in Gurdaspur, Punjab, spans the former floodplains of the Ravi and Beas Rivers, covering 10 km^2. It includes natural marshes, aquaculture ponds, and agricultural wetlands that support crops like lotus and chestnut, surrounded by rice, wheat, and sugarcane fields. With a subtropical, semi-arid climate and an annual rainfall of 959 mm, the reserve hosts over 20,000 waterbirds, including vulnerable species like the sarus crane and woolly-necked stork. The site has a mean eutrophicated and turbid area of 8.78% and 91.48%, respectively. The site's surroundings have a built-up and cropland area of 0.54 and 17.34 km^2, respectively, with an overall wetland health score of 82.2%. The site is recognized under Ramsar Criteria 2 and 5, and it faces threats from invasive species like water hyacinths. The reserve maintains alkaline and freshwater quality and supports local livelihoods.

(23) Khijadia Wildlife Sanctuary

The site, a 5.12 km^2 wetland in Gujarat state, is located at the confluence of the Ruparel and Kalindri Rivers near Jamnagar. Formed by two man-made bunds to retain freshwater, it features saline clay soils influenced by the nearby Gulf of Kutch. The sanctuary experiences a subtropical desert climate with temperatures ranging from 7 to 40 °C. A migratory bird hotspot on the Central Asian Flyway, it supports 312 bird species and meets six Ramsar Criteria (2, 3, 4, 5, 6, 8). The site has a mean eutrophicated and turbid areas of 14.10% and 56.52%, respectively. The site surroundings have a built-up and cropland area of 1.48 and 38.95 km^2, respectively, with an overall wetland health score of 91.1%. The key threats the site faces include water abstraction and chemical runoff. Effluents in the wetland should be monitored properly.

(24) Koonthankulam Bird Sanctuary

The site was declared a Ramsar site in 1994, with an area of 0.72 km^2 in Tamil Nadu's Tirunelveli district. The sanctuary includes the Koonthankulam and Kadankulam irrigation tanks, fed by canals from rivers originating in the Western Ghats. It experiences a tropical humid climate with annual rainfall between 750–850 mm and serves as a sediment sink, supporting groundwater recharge. The site meets Ramsar Criteria 3, 4, 5, and 6, supporting over 1% of South Asian bird populations. The site has a mean eutrophicated and turbid area of 19.79% and 62.04%, respectively. The site surroundings have a built-up and cropland area of 0.14 and 2.43 km^2, respectively, with an overall wetland health score of 84.4%. The key threats site observing include encroachments and shrimp farming, where management should focus on avifauna conservation and local agricultural support.

(25) Loktak Lake

The site spreads over 266 km^2 in Manipur's Bishanpur district. It is Northeast India's largest natural wetland and supports flood control, irrigation, hydropower, and water supply for nearby communities (Criterion 1). The lake is known for its floating mats of vegetation, called 'phumids', and is home to Keibul Lamjao National Park, which protects the endangered brow-antlered deer, or Sangai (Criterion 2). The site has a mean eutrophicated and turbid areas of 27.46% and 3.48%, respectively. The site surroundings have a built-up and cropland area of 5.24 and 102.30 km^2, respectively, with an overall wetland health score of 75.6%. The site observes threats like deforestation, siltation, and invasive species. A government-funded management plan focuses on ecological restoration. Apart from that, community participation should be used to manage the site more efficiently.

(26) Lonar Lake

The site, located in Maharashtra's Buldhana district, is spread over 4.27 km^2 and was formed by a meteorite impact, making it a unique saline and alkaline wetland. Its ecosystem includes 160 bird species and microorganisms adapted to high salinity (Criterion 3). Lonar Lake serves as a vital wintering site for migratory birds along the Central Asian Flyway (Criterion 4) and supports rare species like Indian Sandalwood (Criterion 2) and also Criterion 1. The site has a mean eutrophicated and turbid areas of 43.87% and 35.17%, respectively. The site surroundings have a built-up and cropland area of 1.04 and 6.38 km^2, respectively, with an overall wetland health score of 86.7%. The Maharashtra Forest Department manages key threats, including hydrological imbalances due to climate change. The lake's highly alkaline water exceeds a pH of 10.5, making it unsuitable for most uses.

(27) Longwood Shola Reserve Forest

The site, spanning an area of 1.16 km^2 in Tamil Nadu's Nilgiris district, was designated a Ramsar Site on 24 May 2023. The site is located at 11°26′ 22″ N, 76° 52′ 38″ E, and this tropical montane forest, known as "shola," is interspersed with grasslands. It regulates local microclimates and supports biodiversity, including the endangered Black-chinned Nilgiri Laughing Thrush and Nilgiri Wood-pigeon, meeting Ramsar Criterion 2. It also falls under Criterion 1, 3 and 4. The site has mean eutrophicated and turbid areas of 93.27% and 2.53%, respectively. The site surroundings have a built-up and cropland area of 0.21 and 0.01 km^2, respectively, with an overall wetland health score of 75.6%. The site is facing threats like encroachment and water contamination from nearby tea plantations.

The site management carried out by the Longwood Shola Watchdog Committee helps maintain ecological balance and provides high-quality freshwater.

(28) Magadi Kere Conservation Reserve

The site is spread in 0.54 km^2 in the Shirahatti taluk, Gadag district. The site was originally constructed for irrigation, and increased water salinity reduced its agricultural use, but it became an Important Bird Area. The wetland hosts over 166 bird species, including vulnerable species like the Common Pochard and River Tern, and is a wintering ground for over 8,000 Bar-headed Geese. The site meets Ramsar Criteria 2, 3, 5, and 6. The site has a mean eutrophicated and turbid areas of 21.36% and 20.90%, respectively. The site surroundings have a built-up and cropland area of 0.16 and 1.66 km^2, respectively, with an overall wetland health score of 60%. The increased salinity is a key threat, with conservation efforts focusing on water quality monitoring.

(29) Nalsarovar Wetland

The site is a natural freshwater lake in Gujarat, India, spans 120 km^2, and was designated a Ramsar site on 24 September 2012. The site is located in the Thar Desert Biogeographic Province, it supports 216 bird species, 13 mammals, 11 reptiles, and 19 fish species, including the endangered Indian Wild Ass, Sarus Crane, and Marbled Teal. The site meets Ramsar Criteria 1, 2, 3, 4, 5, and 6 for its size, biodiversity, and support for endangered species. The site has a mean eutrophicated and turbid areas of 30.31% and 18.97%, respectively. The site surroundings have a built-up and cropland area of 0.21 and 19.12 km^2, respectively, with an overall wetland health score of 97.8%. It signifies that the Deputy Conservator of Forests manages the wetland very well.

(30) Nagi Bird Sanctuary

The site is located in Jamui district, Bihar, spread over 2.05 km^2 and centers around a reservoir on the Nagi River. The site was originally established for irrigation; however, it now supports diverse migratory and resident bird species, including 1,600 bar-headed geese, which account for over 1.6% of the flyway population. Endangered species like the black-bellied tern and ferruginous duck are also found here. The site has a mean eutrophicated and turbid areas of 5.95% and 15.08%, respectively. The site surroundings have a built-up and cropland area of 0.14 and 6.48 km^2, respectively, with an overall wetland health score of 82.2%. The site fulfills the Ramsar Criteria 2, 3, 4, 5, and 6, and faces threats from anthropogenic activities and climate change. The site management focuses on habitat conservation and community engagement.

(31) Nakti Bird Sanctuary

The site is a man-made wetland in Jamui district, Bihar, which spans 3.32 km^2 and was formed by the Nakti Dam on the Nakti River. Located 31 km from Jamui Railway

Station, it is an Important Bird and Biodiversity Area (IBA), hosting over 75 bird species, including threatened species like the black-bellied tern and lesser adjutant stork. The reservoir supports over 30,000 waterbirds. The site meets Ramsar Criteria 2, 3, 4, 5, and 6. The site has a mean eutrophicated and turbid areas of 3.07% and 40.91%, respectively. The site surroundings have a built-up and cropland area of 0.03 and 4.72 km^2, respectively, with an overall wetland health score of 82.2%. The site is facing threats like habitat degradation and pollution, with management focusing on pollution control, regulated tourism, and community conservation.

(32) Nanda Lake

The site is a freshwater marsh in Goa, India, spans 0.92 km^2 and holds significant ecological value, meeting Criteria 1, 2, and 3, as it represents a rare wetland type, supports vulnerable and endangered species, and exhibits high biodiversity. The site has a mean eutrophicated and turbid areas of 37.63% and 51.23%, respectively. The site surroundings have a built-up and cropland area of 0.02 and 0.18 km^2, respectively, with an overall wetland health score of 86.7%. Despite threats from urban expansion and pollution, the lake is crucial for agriculture, fishing, and recreation while also protecting surrounding areas from floods.

(33) Nandur Madhameshwar Wetland

The site is in Maharashtra state in an area of 144.41 km^2. This wetland supports biodiversity conservation, meeting several Ramsar Criteria, including support for threatened species (Criterion 2), maintenance of regional biodiversity (Criterion 3), and importance for migratory birds (Criterion 4). It also meets criteria 5, 6, 7, and 8. The site has a mean eutrophicated and turbid areas of 37.88% and 29.82%, respectively. The site surroundings have a built-up and cropland area of 1.88 and 73.47 km^2, respectively, with an overall wetland health score of 84.4%. The site observes threats from human activities and is managed as a Wildlife Sanctuary to protect its ecological integrity. The key management strategies include regular monitoring, habitat restoration, controlling invasive species, and maintaining optimal water quality to preserve the wetland's ecological character.

(34) Nangal Wildlife Sanctuary

The site was designated as a Ramsar Site in 2019, and it is spread over an area of 1.16 km^2 in Punjab state. The site begins from a human-made reservoir and features permanent rivers, streams, and wetlands, providing vital habitat for diverse wildlife and serving as a stopover for migratory birds. It hosts vulnerable and endangered species, including the Greater Spotted Eagle, Indian Pangolin, and Leopard. Meeting Ramsar Criteria 2 and 3 supports biological diversity. The site has a mean eutrophicated and turbid areas of 0.77% and 0.06%, respectively. The site surrounding has a built-up and cropland area of 2.31 and 1.61 km^2, respectively, with an overall wetland health score of 68.9%. The site's key threats include habitat degradation, water pollution, and invasive species. The Department of Forests and Wildlife Preservation manages the site, collaborating with WWF-India.

(35) Nanjarayan Bird Sanctuary Spans

The site is located in Tiruppur District, Tamil Nadu, state, in an area of 1.26 km^2 at coordinates 11° 08′ 03″ N and 77° 23′ 10″ E. The sanctuary is a crucial groundwater recharge zone that supports surrounding agricultural needs. It is home to over 191 bird species, 87 butterfly species, and 21 reptile species, including the Near Threatened Oriental Darter and the Vulnerable Indian Flap-shelled Turtle. Meeting Ramsar Criteria 2, 3, 4, and 6, it maintains biodiversity and serves as a critical habitat during breeding and migration. The site has a mean eutrophicated and turbid areas of 52.23% and 17.62%, respectively. The site surrounding has a built-up and cropland area of 1.38 and 2.32 km^2, respectively, with an overall wetland health score of 68.9%. The Tamil Nadu Forest Department manages the site, focusing on habitat restoration, monitoring, and education programs.

(36) Nawabganj Bird Sanctuary

The site is spread over an area of 2.24 km^2 in the Unnao district of Uttar Pradesh. This shallow marshland is a haven for diverse ecological communities, hosting over 200 plant species, 220 bird species, and various fauna, making it a vital wintering site for migratory birds. Its unique hydrology, influenced by monsoonal rains and the Sarda Canal, supports rare and threatened species like the Egyptian Vulture and Pallas's Fish Eagle. The site has mean eutrophicated and turbid areas of 55.43% and 10.64%, respectively. The site surrounding has a built-up and cropland area of 0.51 and 3.41 km^2, respectively, with an overall wetland health score of 75.6%. The UP Forest and Wildlife Department manages the sanctuary, focusing on conservation and ecological monitoring as the site fulfills Ramsar Criteria 2, 3, 4, 5, and 6, it plays a crucial role in maintaining regional ecological balance.

(37) Pala Wetland

The site is spread over an area of 18.50 km^2 sanctuary in Mizoram state. This vital ecological habitat supports 222 bird species and 227 plant species, including endangered species like the hoolock gibbon and slow loris. The site fulfills the Ramsar Criteria 1, 2, 3, and 4 and is a unique natural wetland that provides critical support during life cycle stages for many species. The site has a mean eutrophicated and turbid areas of 98.39% and 0.28%, respectively. The site's surroundings have a built-up and cropland area of 0.02 and 0.47 km^2, respectively, with an overall wetland health score of 71.1%. The Mara Autonomous District Council and the Department of Environment and Forest manage the sanctuary, focusing on habitat restoration, legal protection, and community engagement to maintain its ecological character and sustain conservation efforts.

(38) Parvati Arga Bird Sanctuary

The site is spread over an area of 7.22 km^2. The Ramsar Site in Uttar Pradesh supports 37 flora species and 64 fauna species. The sanctuary's rich biodiversity is characterized by varied aquatic habitats and floodplain wetlands, making it a critical wintering site for migratory birds and a spawning ground for fish species. Meeting

Ramsar Criteria 2, 3, 4, 5, and 8, the site supports rare and threatened species, maintains biodiversity, and provides essential habitat. The site has a mean eutrophicated and turbid area of 22.82% and 15.90%, respectively. The site's surroundings have a built-up and cropland area of 2.49 and 56.85 km^2, respectively, with an overall wetland health score of 82.2%. The site, managed by the Uttar Pradesh State Wetlands Authority and the Department of Forests, Wildlife, and Climate Change, focuses on habitat conservation, biodiversity monitoring, and sustainable tourism.

(39) Pong Dam Lake

The site, also known as Maharana Pratap Sagar, is a water storage reservoir on the Beas River in Himachal Pradesh, India, covering an area of 156.62 km^2. The site was formed in 1975, as it serves multiple functions, including irrigation and hydropower generation, set against a backdrop of mixed deciduous and pine forests along the Dhaula Dhar Mountain range. The lake is a crucial habitat for over 220 bird species, supporting more than 20,000 migratory waterfowl and notable fish species. It meets Ramsar Criteria 2, 5, and 8, highlighting its ecological significance. The site has a mean eutrophicated and turbid area of 9.12% and 24.67%, respectively. The site's surroundings have a built-up and cropland area of 11.60 and 129.43 km^2, respectively, with an overall wetland health score of 86.7%. The site faces threats from human activities, sedimentation, and agricultural runoff. Management strategies include conservation efforts and habitat monitoring to ensure the lake's ecological health while maintaining its functional roles.

(40) Ranganathittu Bird Sanctuary

The site is located in Karnataka, India, spans 5.177 km^2, and consists of three main islands in the Cauvery River. The site has a mean eutrophicated and turbid area of 31.87% and 8.20%, respectively. The site's surroundings have a built-up and cropland area of 0.03 and 5.24 km^2, respectively, with an overall wetland health score of 73.3%. This unique riverine ecosystem supports a rich biodiversity, featuring 188 plant species, 225 bird species, 69 fish species, and 13 frog species. The sanctuary meets Ramsar Criteria 2, 3, 4, 6, 7, and 8, supporting rare and threatened ecological communities. Notable species include the mugger crocodile and the endangered hump-backed mahseer. A specific management plan needs to be implemented at the site.

(41) Renuka Wetland

The site is in Himachal Pradesh state in an area of 10 km^2 and is a natural lake surrounded by subtropical forests, hosting 443 species of fauna. The wetland qualifies under Ramsar Criteria 3 and 4 due to its rich biodiversity. However, it faces threats from sediment influx, weed growth, and human pressure from six villages housing 2,700 people and 3,407 cattle. The site has a mean eutrophicated and turbid area of 93.27% and 2.53%, respectively. The site's surroundings have a built-up and cropland area of 0.21 and 0.01 km^2, respectively, with an overall wetland health score of 86.7%. The management plan aims to conserve the wetland through retaining walls, plantation drives, and check dams to control soil erosion.

(42) Ropar Wetland

The site, having an area of 13.65 km^2, is a man-made freshwater lake in Punjab state and is a vital ecological zone supporting biodiversity and hydrological functions. The wetland is home to 154 bird species, including rare species like the golden-backed woodpecker and vulnerable species such as the Indian pangolin and smooth Indian otter. Meeting Ramsar Criteria 2 and 3, the site represents a significant habitat for biodiversity, supports vulnerable species, and maintains genetic and ecological diversity. The site has a mean eutrophicated and turbid areas of 18.58% and 38.14%, respectively. The site surrounding has a built-up and cropland area of 1.03 and 16.22 km^2, respectively, with an overall wetland health score of 51.1%. Effective management and protection measures are critical to ensuring sustainability and addressing threats like invasive species and habitat degradation.

(43) Rudrasagar Lake

The site is in West Tripura, India, spans 2.4 km^2, and is a natural freshwater reservoir supporting 443 species of flora and fauna. The lake meets Ramsar Criteria 2, 3, and 8 for hosting endangered species, maintaining biodiversity, and serving as a natural breeding ground for indigenous fish species. The site has a mean eutrophicated and turbid areas of 27.53% and 3.11%, respectively. The site surrounding has a built-up and cropland area of 0.47 and 4.74 km^2, respectively, with an overall wetland health score of 66.7%. However, it faces threats from continuous siltation, increased human habitation, deforestation, and agricultural activities. A Management Action Plan has been submitted to protect the lake, and effective conservation efforts are essential to maintain its ecological health and biodiversity.

(44) Sakhya Sagar

The site is a human-made wetland in Madhya Pradesh that spans 2.48 km^2. Created in 1918 by damming the Manier River, it is integral to the Madhav National Park ecosystem, supporting diverse flora and fauna, including 16 macrophyte species, 19 fish species, 73 bird species, most notably the vulnerable sarus crane, nine reptile species (including marsh crocodiles), and 19 mammal species. The wetland meets Ramsar Criteria 2, 3, and 7 for its ecological significance. The site has a mean eutrophicated and turbid areas of 26.60% and 27.71%, respectively. The site surrounding has a built-up and cropland area of 0.15 and 0.56 km^2, respectively, with an overall wetland health score of 73.3%. The site is facing threats like habitat degradation, pollution, and invasive species like water hyacinths. The Madhya Pradesh Forest Department manages efforts to maintain water levels and preserve ecological integrity.

(45) Saman Bird Sanctuary

The site is in an area of 5.26 km^2 and is a natural wetland in Uttar Pradesh, India. It is a biodiversity hotspot, supporting 187 bird species, including rare and threatened species like the greater spotted eagle and sarus crane. Meeting Ramsar Criteria 2, 3,

4, 5, and 6, the site provides critical habitat for various species, including migratory birds, and maintains biological diversity.

The site has a mean eutrophicated and turbid areas of 29.26% and 58.60%, respectively. The site surrounding has a built-up and cropland area of 0.31 and 6.36 km^2, respectively, with an overall wetland health score of 68.9%. The Forest Department of Uttar Pradesh manages the sanctuary, focusing on maintaining water levels, conserving habitat, and monitoring bird populations. The structured conservation plans are essential to address threats like habitat destruction and water pollution, ensuring the sanctuary's long-term health.

(46) Samaspur Bird Sanctuary

The site is located in the state of Uttar Pradesh in an area of 7.99 km^2. The sanctuary meets several Ramsar Criteria and hosts 149 plant species, 46 fish species, and over 250 bird species, including threatened and endangered species like the sarus crane and greater spotted eagle. It is a biodiversity hotspot with significant seasonal variations in water levels, influenced by monsoon rains and canal inputs. The site meets Criteria 2, 3, 4, 5, 6, and 7. The site has a mean eutrophicated and turbid areas of 34.93% and 32.02%, respectively. The site surrounding has a built-up and cropland area of 0.88 and 35.05 km^2, respectively, with an overall wetland health score of 84.4%. However, it faces threats from habitat destruction, water pollution, and human encroachment. Effective management and conservation efforts are crucial to protect this ecologically significant site and maintain its high biodiversity.

(47) Sambhar Lake

The site is located in the state of Rajasthan in an area of 240 km^2. This shallow wetland meets Ramsar Criteria 1 and is renowned for hosting large numbers of flamingos, waders, and migratory ducks, including rare and threatened species like the desert cat and desert fox. The site has a mean eutrophicated and turbid areas of 5.87% and 67.50%, respectively. The site surrounding has a built-up and cropland area of 23.48 and 766.75 km^2, respectively, with an overall wetland health score of 64.4%. However, the lake faces significant threats from siltation, soil salinization, and sewage discharge from nearby towns. A detailed management plan is being prepared to conserve the wetland through funding from the central government and other agencies. Its international and national importance makes it critical for conservation efforts.

(48) Sandi Bird Sanctuary

The site is in an area of 3.08 km^2 freshwater swamp in Hardoi district, Uttar Pradesh state. The sanctuary hosts over 150 bird species, 13 fish species, 3 amphibians, 15 reptiles, and various butterflies and higher vertebrates. It meets several Ramsar Criteria, i.e., 2, 3, 4, 5, 6, 7, supporting rare and threatened species like the greater spotted eagle and sarus crane while sustaining high biological diversity. The site has a mean eutrophicated and turbid areas of 40.45% and 39.84%, respectively. The site surrounding has a built-up and cropland area of 0.15 and 10.34 km^2, respectively, with an overall wetland health score of 88.9%. The site's key threats include habitat degradation, water pollution, and invasive species. Management focuses on maintaining water levels, habitat restoration, and monitoring bird populations.

(49) Sarsai Nawar Jheel

The site spreads over an area of 1.61 km^2 Ramsar Site in Uttar Pradesh state. This permanent marsh wetland supports a diverse array of species, serving as a critical roosting area for the largest flock of sarus cranes in the region and a significant wintering site for migratory bird species. Meeting Ramsar Criteria 2, 3, 5, 7, and 8, it supports vulnerable species, maintains biological diversity, and provides habitat for riverine fish. The site has a mean eutrophicated and turbid areas of 34.89% and 49.55%, respectively. The site surrounding has a built-up and cropland area of 0.25 and 10.70 km^2, respectively, with an overall wetland health score of 84.4%. The Uttar Pradesh Forest and Wildlife Department manages the site, implementing conservation measures to address threats like agricultural runoff and pollution.

(50) Sasthamkotta Lake

This site is Kerala's largest freshwater lake, spreading over an area of 3.73 km^2 in the Kollam District. It serves as a vital drinking water source and supports diverse fish species. The lake's unique water composition lacks common salts and minerals, inhibiting aquatic plant growth. It qualifies under four Ramsar Criteria: as the largest freshwater lake with unique water quality (Criterion 1), supports endangered species (Criterion 2), hosts significant fish biodiversity (Criterion 7), and provides critical habitat for 27 freshwater fish species (Criterion 8). The site has mean eutrophicated and turbid areas of 42.59% and 16.27%, respectively. The site surrounding has a built-up and cropland area of 0.91 and 0.59 km^2, respectively, with an overall wetland health score of 88.9%. The site's key threats include pollution, agriculture, and urbanization. While no official management plan exists, a proposed action plan awaits funding. The water quality shows neutral pH and low electrical conductivity in both surface and interstitial water.

(51) Satkosia Gorge

The site spreads over an area of 981.97 km^2 in Odisha state. This diverse landscape of marshes and evergreen forests supports a wide array of species, including endangered species like the gharial, mugger crocodile, and greater spotted eagle. It also serves as a breeding ground for commercially important fish species. The site meets Ramsar Criteria 2, 3, 4, 5, and 7, supporting rare species and maintaining ecological balance. The site has a mean eutrophicated and turbid areas of 86.63% and 9.07%, respectively. The site surrounding has a built-up and cropland area of 5.68 and 416.30 km^2, respectively, with an overall wetland health score of 86.7%. The site is managed by the Forest, Environment and Climate Change Department of Odisha, and conservation efforts focus on habitat restoration and sustainable development to address threats such as deforestation, poaching, and water pollution.

(52) Shallabugh Wetland Conservation Reserve

The site is located in the deltaic region of Sindh Nallah, about 18 km west of Srinagar in Jammu and Kashmir, India. It covers an area of 16.75 km^2 at an elevation of 1,580 m. The wetland is part of the Indus Himalayan Foothills and supports diverse flora and fauna, including migratory birds and endemic species. It meets criteria 1, 2, 3, and 4 for its representative wetland type, support of rare species, biological diversity, and crucial role in the life cycles of various species. The site has a mean eutrophicated and turbid areas of 32.94% and 27.67%, respectively. The site surrounding has a built-up and cropland area of 3.23 and 24.56 km^2, respectively, with an overall wetland health score of 62.2%. The wetland faces threats from siltation, eutrophication due to agricultural runoff, habitat loss from encroachment, and climate change. The site management plans focus on habitat enhancement, water quality improvement, and regulating water abstraction.

(53) Sirpur Wetland

The site is located in Indore, Madhya Pradesh, and spreads over an area of 1.61 km^2 with a 10 km^2 catchment area. Located at 22° 41' 58'' N and 75° 48' 44'' E, the wetland, historically constructed by the Holkers, now serves as a crucial water source and supports groundwater recharge. It hosts 6 macrophyte species, 30 fish species, 130 bird species, and 175 terrestrial plant species. Meeting Ramsar Criteria 2, 3, 4, and 7, it supports rare species and maintains high biodiversity. The site has a mean eutrophicated and turbid areas of 45.63% and 8.28%, respectively. The site surrounding has a built-up and cropland area of 2.95 and 1.99 km^2, respectively, with an overall wetland health score of 84.4%. The site faces key threats like pollution, invasive species, and habitat degradation. The Indore City Municipal Corporation manages the site, and conservation focuses on ecological preservation.

(54) Suchindram Theroor Wetland Complex

The site is located in Tamil Nadu state in an area of 0.94 km^2 at an altitude of 4 m. This interconnected system of Suchindram Kulam and Theroor Kulam plays a crucial role in flood control, groundwater recharge, and ecological balance. With a tropical monsoon climate, it supports habitats for various species, hosting over 31,960 waterbirds annually, including *Anhinga melanogaster* and *Channa orientalis*. The site fulfills the Ramsar Criteria 2, 3, 4, 5, 6, and 7, and it supports vulnerable species and critical life stages. The site has a mean eutrophicated and turbid area of 75.12% and 1.77%, respectively. The site's surroundings have a built-up and cropland area of 1.11 and 1.78 km^2, respectively, with an overall wetland health score of 68.9%. The site faces key threats, including urban development, water regulation, and invasive species. The management plan emphasizes habitat restoration, species monitoring, and community engagement.

(55) Sultanpur National Park

The site is located in the Gurugram district, Haryana, in an area of 1.42 km^2. The park is a critical habitat for over 300 bird species, including globally threatened species.

It meets Ramsar Criteria 2 (rare and threatened species), 4 (supporting birds during critical life stages), 5 (supporting over 20,000 waterbirds), and 6 (supporting 1% of waterbird species). The site has a mean eutrophicated and turbid area of 24.33% and 35.13%, respectively. The site's surroundings have a built-up and cropland area of 0.81 and 9.45 km^2, respectively, with an overall wetland health score of 80%. Despite its importance, the park faces threats from human settlements, invasive species, and water pollution. The site is managed by the Forest and Wildlife Department of Haryana, and conservation efforts focus on improving water quality and preserving the park's ecological character.

(56) Surinsar-Mansar Lake

The site is located in Jammu and Kashmir's Udhampur District, in an area of 3.50 km^2. It qualifies as a Ramsar site under Criteria 2 (supporting important turtle species), 3 (hosting rare medusae and diverse flora), and 4 (providing habitat for migratory waterfowl). The lakes are rich in biodiversity, including rare zooplankton taxa. The site has a mean eutrophicated and turbid area of 22.14% and 16.07%, respectively. The site's surroundings have a built-up and cropland area of 0.04 and 0.24 km^2, respectively, with an overall wetland health score of 86.7%. The site is facing threats from pollution, encroachment, and tourism. The site is managed under the Mansar-Surinsar Wildlife Sanctuary, a plan that focuses on scientific research and water quality preservation to maintain ecological balance.

(57) Sur Sarovar Wetland

The site is a man-made reservoir in Uttar Pradesh state, spreads over an area of 4.3 km^2, and is situated 20 km from Agra City. Located in the Yamuna floodplain, it supports 299 plant species, including 24 aquatic plants and over 30,000 waterbirds like herons and sarus cranes. The wetland meets Ramsar Criteria 2, 3, 4, 5, 6, and 7. The site has a mean eutrophicated and turbid area of 44.12% and 9.59%, respectively. The site's surroundings have a built-up and cropland area of 0.37 and 2.41 km^2, respectively, with an overall wetland health score of 84.4%. The site faces key threats, including pollution and invasive species. The site management plan focuses on habitat enhancement and sustainable resource use.

(58) Tawa Reservoir

The site is located in Madhya Pradesh state, and spreads over an area of 200.50 km^2 at the confluence of the Tawa and Denwa Rivers, part of the Narmada Basin. This vital ecological zone supports rich biodiversity, including 57 fish species, over 280 migratory birds, and various plants and reptiles. The reservoir meets Ramsar Criteria 2 (supporting rare species), 3 (maintaining biological diversity), and 4 (providing critical habitats). Additionally, it meets Criteria 7 and 8. The site has a mean eutrophicated and turbid area of 7.81% and 19.35%, respectively. The site's surroundings have a built-up and cropland area of 21.14 and 1093.08 km^2, respectively, with an overall wetland health score of 68.9%. The site faces key threats, including water abstraction, invasive species, and climate change. The Forest Department manages the site, and conservation efforts focus on habitat preservation and water quality improvement.

(59) Thol Lake Wildlife Sanctuary

The site is located in Gujarat state and spreads over an area of 6.99 km^2 with a catchment area of 21.90 km^2. It is situated approximately 25 km from Ahmedabad city, and the sanctuary was originally constructed in 1912 as an irrigation reservoir. It now serves as a crucial habitat for avian biodiversity, supporting 327 bird species, including notable migratory species like the sarus crane and painted stork. Meeting Ramsar Criteria 2, 3, 4, 5, and 6 highlights its ecological significance. The site has mean eutrophicated and turbid areas of 62.62% and 16.41%, respectively. The site's surroundings have a built-up and cropland area of 0.31 and 7.77 km^2, respectively, with an overall wetland health score of 100%. The site's key threats include habitat destruction and pollution. The Forest Department manages the site, and the water quality is alkaline, which is essential for its dual use as drinking and irrigation water.

(60) Tso Kar Wetland Complex

The site is located in the union territory of Ladakh, spreads over an area of 95.77 km^2, and includes the hypersaline Tso Kar and freshwater Startsapuk Tso Lakes. The complex qualifies under four Ramsar criteria: unique wetland type (Criterion 1), supporting rare species (Criterion 2), high biodiversity (Criterion 3), and critical habitat (Criterion 4). It hosts 139 bird species, 232 vascular plants, and endangered mammals like snow leopards. The site has a mean eutrophicated and turbid areas of 3.13% and 66.75%, respectively. The site surrounding has a built-up area of 0.15 km^2 and negligible cropland with an overall wetland health score of 53.3%. The site's key threats include human settlements, tourism, and pollution. The Department of Wildlife Protection manages the site; it plays a crucial role in maintaining the regional water table.

(51) Tsomoriri Wetland

The site spreads over an area of 120 km^2 in Eastern Ladakh as one of the highest lakes in the world, supporting a rich diversity of flora and fauna. It serves as a critical breeding ground for the bar-headed goose and black-necked crane, hosting over 40 species of waterbirds. The site fulfills the Ramsar Criteria 1, 2, and 4. It is a unique natural wetland that provides habitat for numerous waterbird species. The site has a mean eutrophicated and turbid areas of 9.47% and 20.86%, respectively. The site surrounding has a built-up area of 0.16 km^2 and negligible cropland with an overall wetland health score of 80%. The site faces threats such as habitat degradation, pollution, and climate change. Effective management is needed, including regulating tourism and promoting sustainable practices, to preserve its ecological integrity.

(62) Udhayamarthandapuram Bird Sanctuary

The site is located in Thiruvarur District, Tamil Nadu, covering approximately 0.43 km^2 of area. The sanctuary features diverse habitats supported by rainfall and runoff from the Mettur Dam, characterized by a tropical savanna climate with a winter dry season. It hosts over 20,000 waterbirds, including the Near Threatened Black-headed Ibis and Glossy Ibis, and meets Ramsar Criteria 2, 3, 4, and 5. The site has mean eutrophicated and turbid areas of 75.64% and 18.45%, respectively. The site surrounding has a built-up and cropland area of 0.02 and 0.94 km^2, respectively, with an overall wetland health score of 73.3%. The site's key threats include human encroachment, water regulation issues, and invasive species. The District Forest Officer manages the sanctuary, emphasizing sustainable practices for local agricultural communities.

(63) Upper Ganga River

The site is located between the Brijghat and Narora Stretch, spreads over an area of 9238.52 km^2, and features shallow waters, deep pools, and sandy, muddy banks. The site has a presence near districts like Ghaziabad, Bulandshahr, Badaun, and Moradabad, and it experiences three seasons and fluctuating water levels due to irregular hydrology. The wetland supports rich biodiversity, including endangered species like the Ganges River dolphin and *Gavialis gangeticus*, alongside over 100 bird species and 82 fish species. It meets Ramsar Criteria 2, 3, 4, 5, and 7 for supporting endangered species, maintaining biodiversity, and providing habitats for migratory birds. The site has a mean eutrophicated and turbid areas of 18.23% and 55.25%, respectively. The site surrounding has a built-up and cropland area of 2077.41 and 24,818.93 km^2, respectively, with a n overall wetland health score of 66.7%. The site's key threats include pollution, agricultural runoff, and human activities. The conservation efforts by WWF-India focus on pollution control, habitat restoration, and community awareness.

(64) Vaduvur Bird Sanctuary

The site is located in Tamil Nadu's Thiruvarur district, in an area of 1.12 km^2 with a catchment area of 27.80 km^2. The Mettur Dam feeds this human-made irrigation tank through local canals. Situated in the Cauvery Delta, it attracts migratory birds due to its rich biodiversity. The sanctuary experiences a semi-arid climate with an average annual rainfall of 1,129 mm and supports over 118 bird species, including the spot-billed pelican and black-headed ibis. It meets Ramsar Criteria 2 (rare species), 3 (biodiversity), 4 (critical life stages), 5 (over 20,000 waterbirds), and 6 (1% population of certain waterbirds). The site has a mean eutrophicated and turbid areas of 70.53% and 11.08%, respectively. The site surrounding has a built-up and cropland area of 0.07 and 1.61 km^2, respectively, with an overall wetland health score of 77.8%. The Tamil Nadu Forest Department manages the site and faces threats from urbanization and invasive species.

(65) Vellode Bird Sanctuary

The site is located in Tamil Nadu state in an area of 0.77 km^2 with a catchment area of 1.02 km^2. The site has a mean eutrophicated and turbid areas of 63.35% and 11.41%, respectively. The site surrounding has a built-up and cropland area of 0.12 and 1.01 km^2, respectively, with an overall wetland health score of 71.1%. This critical breeding habitat for wetland birds experiences a tropical savanna climate, with elevations ranging from 140 to 160 m above sea level. The sanctuary supports diverse flora and fauna, including vulnerable plant species and various birds of the Central Asian Flyway. It meets Ramsar Criteria 2, 3, and 5. The sites facing key threats include human settlements and invasive species like *Cyprinus carpio*.

(66) Vembannur Wetland Complex

The site spreads over an area of 0.2 km^2 in Kanyakumari District, Tamil Nadu, and is a human-made inland tank located near the Bay of Bengal. The wetland receives water from the Kodaiyar basin and supports around 250 bird species, including *Anhinga melanogaster* and *Pelecanus philippensis*. It meets Ramsar Criteria 2 (rare species), 3 (biological diversity), 4 (critical life stages), and 6 (over 1% of the waterbird population). The site has mean eutrophicated and turbid areas of 79.90% and 0.66%, respectively. The site surroundings have a cropland area of 0.18 km^2 with an overall wetland health score of 73.3% and negligible built-up land cover. The site faces key threats, including urbanization, agriculture, and invasive species like water hyacinths. The management plan focuses on promoting livelihoods and enhancing heronry bird populations.

(67) Wadhvana Wetland

The site is spread over an area of 5.79 km^2, is a man-made wetland in Gujarat state, and has a catchment area of 11 km^2. This century-old irrigation reservoir supports agriculture and biodiversity, hosting around 200 avian species, including over 140 waterbird species. It is a critical habitat for globally threatened species like the Indian Sarus Crane and Black-bellied Tern. The wetland meets Ramsar Criteria 2, 3, 4, 5, 6, and 7. The site has a mean eutrophicated and turbid areas of 12.41% and 10.23%, respectively. The site surrounding has a built-up and cropland area of 0.23 and 12.02 km^2, respectively, with an overall wetland health score of 95.6%. The site's key threats include tourism, urban development, and agriculture. The Gujarat Forest Department manages the site, focusing on controlling invasive plants and preserving ecological integrity.

(68) Wular Lake

It is the largest freshwater wetland site spread over an area of 189 km^2 in Jammu and Kashmir, India, surrounded by high mountain ranges and fed by the River Jhelum. The lake supports a large population of migratory and resident birds, serving as a significant revenue source for the State Government. The site has mean eutrophicated and turbid areas of 0.60% and 0.01%, respectively. The site surrounding has a built-up and cropland area of 14.18 and 79.17 km^2, respectively, with an overall wetland health score of 77.8%. The site fulfills the Ramsar Criteria 1 (unique wetland type), it faces threats from siltation, habitat conversion, and pollution. An Action Plan is in place to address conservation and ecological concerns.

(69) Yashwant Sagar

The site is located on the Gambhir River near Indore, Madhya Pradesh, and spreads over an area of 8.23 km^2. The site is characterized by a tropical climate and annual rainfall of about 1,000 mm. It supports diverse habitats essential for ecological balance. It is recognized as an Important Bird Area (IBA), as it hosts threatened species like the ferruginous pochard and over 20,000 waterbirds, meeting Ramsar Criteria 2, 3, 4, 5, 7, and 8. The site has a mean eutrophicated and turbid areas of 47.41% and 8.28%, respectively. The site surrounding has a built-up and cropland area of 0.61 and 38.63 km^2, respectively, with an overall wetland health score of 66.7%. The site faces key threats, including pollution and invasive species. Management involves local authorities, with a comprehensive plan under development.

4.2.2 Anthropogenic Pollution and Land Degradation

Anthropogenic activities are the primary cause of wetland ecological changes among Inland wetlands. The pollution generated by the agricultural, industrial, and urban sectors, as well as the land reclamation of the wetlands, is discussed in this section.

Runoff Pollution from Agriculture

The agricultural runoff is the most common cause of pollution in the wetlands. The fertilizer and pesticide runoff from agriculture fields in the wetlands contribute to nutrient loading and enhance algal blooms. For Example, Kabartal Wetland [10] and Nandur Madhameshwar [11], face this problem and would be harsh for wetlands, especially with intensive agricultural practices surrounding it. Nutrient pollution through agricultural runoff leads to eutrophication, followed by oxygen depletion that causes a decline in fish population and the destruction of aquatic habitats [12]. The input of excessive nutrients from surrounding fields causes recurrent algal blooms within the wetlands; therefore, it results in poor water quality, making these wetlands less favorable for aquatic life and water birds.

Industrial Pollution

Industries polluting wetlands, like Bhoj Wetland and Wular Lake, are highly vulnerable to such activities. The untreated sewage and industrial waste discharged into the lower lake have continued to enhance eutrophication in Bhoj Wetland [13]. It leads to a decline in the regional water quality supply to the city and impacts biodiversity. Similarly, the existence of industrial activities around Wular Lake results in toxic pollution to its aquatic ecosystem and also impacts local communities relying on it for livelihood purposes [14]. Toxic metals and chemicals from industries disturb natural wetland processes and degrade their ability to filter water and enhance biodiversity.

Urbanization and Wastewater Disposal

Urban encroachment and, subsequently, the disposal of untreated wastewater into wetlands contribute immensely to water quality deterioration. The boom of growth of the adjacent urban territories has impacted some wetlands like Bhoj Wetland [13]. Wastewater disposal, mainly in the form of household sewage and industrial effluents, augmented the concentration of nitrogen, phosphorus, and heavy metals within water bodies. It accelerates eutrophication and impacts the health of the species that depend on these ecosystems.

Land Degradation

Land degradation in wetland surroundings for agriculture, infrastructure development, and urban expansion is another major threat to these wetlands. The conversion of wetland areas to agricultural fields or urban spaces decreases natural habitats and hydrological cycles. Extensive agriculture around the Parvati Arga has changed the hydrology of the water bodies, which leads to a reduced storage capacity and increased sedimentation [15]. In the case of Deepor Beel, the extensive human settlements and urbanization have caused significant encroachment, disrupted natural water flow, and altered its natural ecological functions [16].

4.2.3 Comparison with Ramsar Standards and Global Benchmarks

It is observed that several Ramsar inland wetlands in the nation are lacking in fulfilling all the Ramsar Standards and Global Benchmarks for wetland conservation and community development.

Water Quality Standards

Ramsar promotes maintaining water quality, supporting aquatic life, and minimizing the impacts of eutrophication and pollution. Water quality standards are deficient in many of the wetlands analyzed, such as Bhindawas Wetland. Globally, wetlands like the Everglades in the U.S. have enforced rigorous control measures for pollution to mitigate nutrient loading, thereby establishing norms for wetland management [17]. Indian wetlands, specifically urban wetlands, must strengthen their control measures on pollution to exhibit consistency more often with these standards laid by Ramsar.

Pollution Control and Management

Ramsar standards encourage minimum pollution, waste treatment, and agricultural runoff. Most wetlands have integrated elaborate waste management systems into their system, involving natural wetland processes for the treatment of sewage to maintain water quality. However, Inland Ramsar wetlands are yet to improve the proper management of pollution, with more stringent industrial regulations and improved best practices in agriculture being necessary steps forward. Internationally, wetlands in other countries, for example, Olentangy River Wetland Research Park in the USA, have been able to control pollution through holistic regulatory policies and an active sense of community involvement. Wetlands of Indian origin could learn from such policies to curtail industrial effluents as well as agricultural runoff.

Management of Invasive Species

Ramsar guidelines stress the issue of the "Control measures of invasive species" to preserve the native biodiversity. Although many wetlands in India, such as Harike Lake and Kanjli wetland, are quite penetrated by invasive species like water hyacinths, Koonthankulam Bird Sanctuary is an example of effective control measures against invasive plants against which Prosopis juliflora. Hence, global wetlands, like the Danube Delta, must adopt more efficient methods for removing invasive species so that these become models for Indian wetlands [18].

Biodiversity Conservation

The site management pertaining to Ramsar's effort to protect wetland biodiversity is well reflected in many Indian wetlands, especially those like Nandur Madhmeshwar Wetland, Asan Conservation Reserve, and Keoladeo National Park. Such wetlands demonstrate clear goals of sustenance of migratory bird populations and endangered species, much like global benchmarks observed in wetlands. Kanjli Wetland and Parvati Arga Bird Sanctuary need enhancement actions for biodiversity conservation.

4.3   Case Studies and Innovations in Wetland Management

In order to manage the wetland ecosystem, new innovative wetland management practices are required to restore water quality and minimize the impact of climate change and anthropogenic activities. The details regarding the role of technology and unique challenges for inland wetlands with innovative solutions in wetland management practices are discussed in this section.

4.3.1   Case Studies of Inland Wetland Management Practices

In this section, we present the case studies of two inland wetland management practices, which are presented as follows:

A. Vellode Bird Sanctuary Using Aquatic Insects as Bioindicators

The Vellode Bird Sanctuary in Tamil Nadu is one of the identified important sites for migratory birds. It is supported by Vellode Lake, a man-made tank originally constructed for irrigation purposes. The study by Malathi et al. [19] assesses the ecological quality of the lake using aquatic insects as bioindicators. Aquatic insects are great indicators of wetland health due to their sensitivity to pollution and changes in the environment. The study reported that the site consisted of aquatic insects in 29 families of six orders; *Ephemeroptera* (mayflies) appeared mostly in winter months, implying that the water was cleaner and better oxygenated. *Chironomidae* (non-biting midges) are tolerant of pollution and were most commonly found to appear in lower water quality and higher nutrient sections. Seasonal fluctuations affected the insects' diversity. The highest diversity of insects occurred in the post-monsoon, and the lowest diversity of insects occurred during the monsoon, with probable reasons like increased organic runoff and changes in water levels. The greatest Shannon–Wiener Diversity Index was shown in winter, but the dominance index was greater during the monsoon, which represents a pollution effect on community structure. It has been of high significance because the invertebrates, particularly aquatic insects such as *Ephemeroptera* and *Chironomidae*, represented good indicator organisms of the lake's ecological condition, with sensitive species reflecting the good quality of water and tolerant species indicating the areas of concern. Water quality monitoring by assessing the whole spectrum of data remains vital as a means of controlling pollution within the sanctuary and ensuring the long-term conservation of this wetland ecosystem. Thus, by including bioindicators such as aquatic insects in the management of wetlands, Vellode Bird Sanctuary is protected from degradation since its status will be preserved as a habitat for different species of animals.

B. Phytoremediation in Loktak Lake, Manipur

The Loktak Lake, the largest freshwater lake in Northeast India and a designated Ramsar site, is mainly impacted by iron (Fe) pollution. This iron contamination is mainly from urban waste and agricultural runoff into the lake through inflows from Nambol and Yangoi Rivers. The high concentration of iron is not only dangerous for aquatic ecosystems but for human health as well; thus, the remediation process also needs to be done urgently. The study by Meitei and Prasad [20] explores the potential of four invasive aquatic plant species that can be utilized for phytoremediation: *Pistia stratiotes* (water lettuce), *Lemna minor* (duckweed), *Eichhornia crassipes* (water hyacinth), and *Salvinia cucullata* (floating fern). Among those, the ability of *Pistia stratiotes* to accumulate iron was the highest, and therefore, it is identified as an efficient bio-accumulator. The other species also showed some capacity to absorb iron, making them a potential tool in the mitigation of iron pollution at Loktak Lake. The selected species can thus be utilized for phytoremediation, and in controlling their growth, one can deal with the problem of pollution simultaneously. Overall, the results indicate their role in water quality improvement and show their ability to control invasive populations. In conclusion, the phytoremediation strategy can be used to manage iron contamination in Loktak Lake sustainably. Based on the utility of the bioaccumulation ability of the considered species, the ecological health of the lake can be boosted to enhance community development that substantially relies on its resources. Therefore, this knowledge of efforts toward phytoremediation with conservation is important in the maintenance of the lake's ecological integrity.

4.3.2 Challenges and Solutions in Water Quality Management of Wetlands

The inland Wetlands face numerous challenges that impact water quality and overall ecosystem health. Factors such as agricultural runoff, urbanization, pollution, and invasive species contribute significantly to the deterioration of wetland ecosystems. In this section, we discussed the few challenges they faced, along with solutions that have been implemented to manage water quality effectively.

Challenges in Water Quality Management

Inland wetlands are crucial ecosystems that provide services such as water filtration, carbon sequestration, and biodiversity support. However, they are highly vulnerable to anthropogenic influences, especially from agriculture, urbanization, and climate change [21].

Agricultural Runoff and Eutrophication: Agricultural activities introduce excess nutrients like nitrogen and phosphorus into wetlands through runoff, leading to eutrophication [22]. This process stimulates excessive growth of algae, which eventually depletes oxygen levels in the water, affecting aquatic life. It is a significant issue for inland wetlands such as Harike Wetlands in Punjab state [23], India, where nutrient overload from surrounding agricultural lands has caused detrimental algal blooms and the spread of invasive species. Pesticides and herbicides used in farming further exacerbate the problem by introducing toxic substances that harm biodiversity.

Urbanization and Pollution: Urbanization adds to the degradation of inland wetlands through increased stormwater runoff, which carries pollutants such as heavy metals, oil, microplastics, and untreated sewage into wetland ecosystems [24]. For example, urban wetlands like Naini Lake in Delhi [25] and those in the Bhoj wetland [13] have been significantly affected by urban runoff and untreated sewage inflows. It results in poor water quality, oxygen depletion, and loss of biodiversity, often triggering algal blooms and hypoxic conditions.

Water Extraction and Climate Change: A challenge of water extraction to utilize it for agricultural and domestic purposes, which lowers the water table and reduces wetland size. It diminishes the wetland's ability to filter pollutants naturally. Compounding the issue is climate change, which exacerbates droughts, reduces water availability, and disrupts the natural hydrological cycles that wetlands rely on for functioning efficiently [26].

Design and Maintenance Issues: Artificial wetlands, such as those used for wastewater treatment, often suffer from poor design and inadequate maintenance. When water flow is not adequately maintained, wetlands can become stagnant, leading to the accumulation of pollutants that are not sufficiently diluted or processed [27]. For example, stagnant water allows for the build-up of excess nutrients and contaminants, which can lead to eutrophication and reduce the wetland's efficiency in treating wastewater.

Introduction of Complex Pollutants: Wetlands, particularly those designed for wastewater treatment, often have to manage pollutants not typically found in natural wetlands. These include pharmaceuticals, industrial chemicals, and heavy metals [28]. These contaminants are difficult to break down and pose a significant challenge to nutrient cycling and water purification processes.

Solutions for Water Quality Management

Despite these challenges, several solutions have been developed to mitigate the negative effects of pollution and ensure the sustainability of both inland and artificial wetlands. Effective water quality management relies on integrated approaches that combine sustainable agricultural practices, advanced filtration technologies, and the restoration of natural processes.

Sustainable Agricultural Practices: Reducing agricultural runoff into wetlands can be achieved through the adoption of precision farming, which minimizes the use of fertilizers and pesticides. Organic farming practices can further reduce the toxic chemical load entering wetlands. In some wetlands, promoting organic farming in surrounding areas can reduce the nutrient inflow, thereby mitigating eutrophication and protecting biodiversity [29].

Buffer Zones and Riparian Vegetation: Establishing buffer zones with dense riparian vegetation around wetlands acts as a natural filter, trapping sediments, nutrients, and pollutants before they reach the wetland [30]. These vegetative buffers slow down water runoff, allowing contaminants to settle out of the water before entering the ecosystem.

Water-Sensitive Urban Design (WSUD): In urban areas, WSUD strategies such as the construction of swales, detention basins, and constructed wetlands are highly effective. These methods can capture and treat stormwater before it reaches natural wetlands, significantly reducing the pollution load [31].

Proper Design and Maintenance: To prevent issues like water stagnation, artificial wetlands need to be carefully designed to ensure consistent water flow. Regular monitoring and maintenance of the system are critical to ensuring optimal performance in pollutant removal. Adequate water flow reduces the risk of pollutant accumulation and enhances the efficiency of nutrient uptake by wetland plants [32].

Constructed Floating Wetlands (CFWs): CFWs are floating platforms planted with wetland species that can uptake excess nutrients such as nitrogen and phosphorus from the water. These floating wetlands act as living filtration systems and are particularly effective in wastewater treatment settings. By using plants with high nutrient absorption rates, CFWs help reduce the levels of pollutants while also enhancing habitat quality for aquatic species [33].

Advanced Filtration Technologies: Artificial wetlands benefit from integrating advanced filtration systems like biofilters and membrane technologies. These technologies enhance the removal of specific contaminants [34] such as pharmaceuticals, heavy metals, and other industrial pollutants that are challenging to treat through natural processes alone [35].

Bioremediation and Phytoremediation: The use of wetland plants and microorganisms to break down contaminants is a promising solution for both inland and artificial wetlands. This process, known as bioremediation, utilizes the natural ability of plants and microbes to absorb and detoxify harmful substances [36]. Wetland plants, such as reed beds, have been successfully used to treat pollutants in wastewater and enhance water quality in both types of wetlands [37]. Water quality management in both inland and artificial wetlands presents a complex set of challenges, including nutrient overload, toxic pollutants, and the impacts of urbanization and climate change. However, sustainable agricultural practices, water-sensitive urban design, and advancements in constructed wetland technology offer viable solutions to mitigate these challenges. By improving design, monitoring systems, and restoration efforts, wetland ecosystems can continue to provide essential services like water filtration, biodiversity support, and climate regulation. These strategies are crucial for protecting wetlands and enhancing their resilience to future environmental pressures.

4.3.3 Implications to Ecosystem Health and Biodiversity

Wetlands are significant ecosystems that offer essential ecosystem services such as water filtering, carbon sequestration, and habitat provision. The disturbances in water quality, both inland and artificial wetlands, threaten biodiversity and ecosystem health significantly. Inland wetlands are a critical type of habitat for many species, such as fish, amphibians, birds, and invertebrates. The lack of good water quality induces biodiversity loss caused by the sensitivities of many species and the changes in nutrient levels and toxins. For instance, eutrophication leads to hypoxia, which is a situation of a lack of adequate oxygen levels needed for the survival of aquatic life, leaving many fish and amphibians dead. Bioaccumulation can also occur due to the presence of heavy metals and pesticides within the water. It affects the species further up in the food chain, such as birds and mammals, specifically migratory birds that depend on wetlands as a source of breeding and feeding. Loss of biodiversity causes a shift in the ecological equilibrium of wetland ecosystems, resulting in species composition, which decreases the healthiness of the ecosystem as a whole. For example, if more tolerant invasive species can outcompete native species, their introduction will further degrade biodiversity. In addition to their contribution to conserving biodiversity, artificial wetlands help treat wastewater and manage stormwater. However, the sustainable operation of artificial wetlands for maintaining ecosystem health still lies in their management. If inadequately managed, artificial wetlands become ecological traps: species are drawn to the area but find that the water quality is atrocious, thus lowering survival and reproductive rates. However, properly managed artificial wetlands will make positive contributions to biodiversity. Habitat provided in areas where natural wetlands have been lost or degraded may support species diversity and enhance ecosystem services such as water filtration and flood control [12]. Artificial wetlands, properly designed, will support a range of plant and animal species and serve as valuable conservation tools for biodiversity in urban and agricultural landscapes.

4.3.4 Role of Technology in Improving Wetland Management

Besides, technological innovations have brought about a tremendous change in wetland management, especially for artificial and inland wetlands. The quality of water and biodiversity support cannot be maintained without the advancement of restoration, and pollution monitoring and control technologies [38]. Therefore, in this section, we discuss the monitoring and restoration technologies for wetlands.

Monitoring Technologies

There has been a dramatic improvement in monitoring wetland management using remote sensing and GIS [39]. The health of wetlands can be monitored continuously by tracking changes in water level, vegetation, and movement of pollutants. Large wetland areas can be observed using satellite imagery and drones, thus not disturbing the ecosystem and bringing valuable data. It will help in the implementation of more effective management plans and the earlier detection of issues before they become critical [40].

Restoration Technologies

Restoration technologies for degraded wetlands are becoming increasingly important. Examples of these techniques include hydrological restoration, a measure aimed at re-establishing natural water flows in inland wetlands as a way of rehabilitating these wetlands. Plant-based technologies refer to the way, for example, wetland plants are used in phytoremediation to uptake and break pollutants, thus rejuvenating contaminated wetlands back to a state of nature. Constructed wetland technology imitates the natural functions of wetlands like artificial wetlands [41]. These systems treat stormwater and wastewater, eliminating contaminants through biological and physical processes. Subsurface flow systems assist in reducing contamination, and improvement in pollutant uptake efficiency is now increasingly achieved through bioengineered plants.

Smart Water Management Systems

Smart technology, including automated water control systems, now exists in the management of wetlands. Real-time monitoring of water levels, pollution loads, and ecosystem conditions is achievable with such a setup, thus allowing for better adjustment of management practices when necessary. For example, real-time detection of changes by smart sensors can cause quick interventions before ecosystem degradation may occur. In artificially built wetlands, smart water systems would ensure efficient treatment of wastewater through adaptation based on real-time data of flow and aeration rates [42]. Thus, stagnation conditions would be reduced considerably, and pollutants would be removed before the water is released into the environment [43]. Considering this, in the subsequent section, we discussed the strategic action plan for wetland ecosystem restoration and its monitoring techniques in detail.

References

1. US EPA, *What Is a Wetland?* (US EPA, 2024), https://www.epa.gov/wetlands/what-wetland#:~:text=Inlandwetlandsincludemarshesand,Typesforafulllist
2. Ramsar, *Ramsar Sites Information Service* (Ramsar, 2024), https://rsis.ramsar.org/ris-search/
3. N. Bassi, M.D. Kumar, A. Sharma, P. Pardha-Saradhi, Status of wetlands in India: a review of extent, ecosystem benefits, threats and management strategies. J. Hydrol. Reg. Stud. **2**, 1–19 (2014). https://doi.org/10.1016/j.ejrh.2014.07.001
4. Ramsar, *Urban Wetlands: Prized Land, Not Wasteland* (Ramsar, 2024), https://www.ramsar.org/sites/default/files/documents/library/wwd18_handouts_eng.pdf
5. L. Venkatachalam, *Economic Valuation of Wetland Ecosystem Services* (2023), https://www.mids.ac.in/assets/doc/WP_242.pdf
6. M.K. Goyal, C.S.P. Ojha, D.H. Burn, Nonparametric statistical downscaling of temperature, precipitation, and evaporation in a semiarid region in India. J. Hydrol. Eng. **17**(5), 615–627 (2012)
7. A. Azhoni, M.K. Goyal, Diagnosing climate change impacts and identifying adaptation strategies by involving key stakeholder organisations and farmers in Sikkim, India: Challenges and opportunities. Sci. Total. Environ. **626**, 468–477 (2018)
8. Y. Xi, S. Peng, P. Ciais, Y. Chen, Future impacts of climate change on inland Ramsar wetlands. Nat. Clim. Change **11**(1), 45–51 (2021). https://doi.org/10.1038/s41558-020-00942-2
9. SACON, *Drained India's Inland Wetlands. Down to Earth* (2022), https://www.downtoearth.org.in/environment/drained-indias-inland-wetlands-9258
10. G. Chaturvedi, K. Avishek, Geospatial approach to identify the indicators of wetland change: a study for Kabartal (Ramsar Wetland), India. Results Eng. 102999 (2024)
11. P. Wagh, S. Kurhade, Monitoring water quality of Nandur Madhmeshwar Wetland, Nasik, Maharashtra (India). In *Proceeding of International Conference SWRDM* (2012), pp. 79–80
12. V.C. Kapsalis, I.K. Kalavrouziotis, Eutrophication—a worldwide water quality issue, in *Chemical Lake Restoration* (Springer International Publishing, 2021), pp. 1–21. https://doi.org/10.1007/978-3-030-76380-0_1
13. C. Sadananda, L.B. Singha, K. Premkumar, P. Lulloo, Comparison of possible impacts on the environment by the organic and conventional farming system in Bhoj Wetlands. Environ. Sustain. **I**, 94 (2022)
14. S. Gupta, Y. Nazir, *Kashmir's Wular Lake Is in Crisis* (Pulitzer Center, 2023). https://pulitzercenter.org/stories/kashmirs-wular-lake-crisis
15. KBA, Parvati Aranga Wildlife Sanctuary, (2024) . https://www.keybiodiversityareas.org/site/factsheet/18428#:~:text=According to the Forest Department,of the Sanctuary are required
16. Nazimuddin Siddique, Encroachment of Deepor Beel threatens livelihood of 1,200 families in Assam (2020)
17. C. Armstrong, T. Piccone, J. Dombrowski, The largest constructed treatment wetland project in the world: the story of the Everglades stormwater treatment areas. Ecol. Eng. **193**, 107005 (2023). https://doi.org/10.1016/j.ecoleng.2023.107005
18. N. Ferreira-Rodríguez, A.J. Castro, B.N. Tweedy, C. Quintas-Soriano, C.C. Vaughn, Mercury consumption and human health: linking pollution and social risk perception in the southeastern United States. J. Environ. Manag. **282**, 111528 (2021)

19. P. Malathi, A. Christine, K. Sivasankaran, A. Manimozhi, *Aquatic Insects: Indicators of Ecological Quality of Waters of Wetland in Tamil Nadu*. Technical report (Advanced Institute for Wildlife Conservation, Tamil Nadu, 2024)
20. M.D. Meitei, M.N.V. Prasad, Bioaccumulation of nutrients and metals in sediment, water, and phoomdi from Loktak Lake (Ramsar site), northeast India: phytoremediation options and risk assessment. Environ. Monit. Assess. **188**(6), 329 (2016). https://doi.org/10.1007/s10661-016-5339-7
21. V. Jain, K.S. Rautela, M.K. Goyal, Ecological restoration: an overview of science and policy regime, in *Ecosystem Restoration: Towards Sustainability and Resilient Development* (2023), pp. 1–27
22. K. Kill, L. Grinberga, J. Koskiaho, Ü. Mander, O. Wahlroos, D. Lauva, J. Pärn, K. Kasak, Phosphorus removal efficiency by in-stream constructed wetlands treating agricultural runoff: influence of vegetation and design. Ecol. Eng. **180**, 106664 (2022). https://doi.org/10.1016/j.ecoleng.2022.106664
23. S. Singh, A. Bhardwaj, V.K. Verma, Remote sensing and GIS based analysis of temporal land use/land cover and water quality changes in Harike wetland ecosystem, Punjab, India. J. Environ. Manag. **262**, 110355 (2020). https://doi.org/10.1016/j.jenvman.2020.110355
24. G.A. Asongwe, I.B. Bame, L.M. Ndam, A.E. Orock, V.A. Tellen, K.P. Bumtu, A.S. Tening, Influence of urbanisation on phytodiversity and some soil properties in riverine wetlands of Bamenda municipality, Cameroon. Sci. Rep. **12**(1), 19766 (2022). https://doi.org/10.1038/s41598-022-23278-7
25. N. Singh, A. Pandey, V. Singh, Impact of institutional overlapping on water governance of Himalayan City, Nainital, India. Water Policy wp2024285 (2024)
26. R. Singh, V. Saritha, C.B. Pande, Monitoring of wetland turbidity using multi-temporal Landsat-8 and Landsat-9 satellite imagery in the Bisalpur wetland, Rajasthan, India. Environ. Res. **241**, 117638 (2024). https://doi.org/10.1016/j.envres.2023.117638
27. L.A. Nuamah, Y. Li, Y. Pu, A.S. Nwankwegu, Z. Haikuo, E. Norgbey, P. Banahene, R. Bofah-Buoh, Constructed wetlands, status, progress, and challenges. The need for critical operational reassessment for a cleaner productive ecosystem. J. Clean. Prod. **269**, 122340 (2020). https://doi.org/10.1016/j.jclepro.2020.122340
28. R.O.B. Makopondo, L.K. Rotich, C.G. Kamau, Potential use and challenges of constructed wetlands for wastewater treatment and conservation in game lodges and resorts in Kenya. Sci. World J. **2020**, 1–9 (2020). https://doi.org/10.1155/2020/9184192
29. A.G. Firth, B.H. Baker, J.P. Brooks, R. Smith, R.B. Iglay, J. Brian Davis, Low external input sustainable agriculture: winter flooding in rice fields increases bird use, fecal matter and soil health, reducing fertilizer requirements. Agr. Ecosyst. Environ. **300**, 106962 (2020). https://doi.org/10.1016/j.agee.2020.106962
30. C.R. Walton, D. Zak, J. Audet, R.J. Petersen, J. Lange, C. Oehmke, W. Wichtmann, J. Kreyling, M. Grygoruk, E. Jabłońska, W. Kotowski, M.M. Wiśniewska, R. Ziegler, C.C. Hoffmann, Wetland buffer zones for nitrogen and phosphorus retention: impacts of soil type, hydrology and vegetation. Sci. Total Environ. **727**, 138709 (2020). https://doi.org/10.1016/j.scitotenv.2020.138709
31. A. Hoban, Water sensitive urban design approaches and their description, in *Approaches to Water Sensitive Urban Design* (Elsevier, 2019), pp. 25–47
32. R.M. Gersberg, S.R. Lyon, R. Brenner, B.V. Elkins, Integrated wastewater treatment using artificial wetlands: a gravel marsh case study, in *Constructed Wetlands for Wastewater Treatment* (CRC Press, 2020), pp. 145–152
33. M. Arslan, S. Wilkinson, M.A. Naeth, M. Gamal El-Din, Z. Khokhar, C. Walker, T. Lucke, Performance of constructed floating wetlands in a cold climate waste stabilization pond. Sci. Total Environ. **880**, 163115 (2023). https://doi.org/10.1016/j.scitotenv.2023.163115
34. M.K. Goyal, S. Kumar, A. Gupta, *AI for Water Policy* (2024), pp. 41–53. https://doi.org/10.1007/978-3-031-72014-7_4
35. S. Deng, J. Chen, J. Chang, Application of biochar as an innovative substrate in constructed wetlands/biofilters for wastewater treatment: performance and ecological benefits. J. Clean. Prod. **293**, 126156 (2021). https://doi.org/10.1016/j.jclepro.2021.126156

36. L. Chen, J. Beiyuan, W. Hu, Z. Zhang, C. Duan, Q. Cui, X. Zhu, H. He, X. Huang, L. Fang, Phytoremediation of potentially toxic elements (PTEs) contaminated soils using alfalfa (Medicago sativa L.): a comprehensive review. Chemosphere **293**, 133577 (2022). https://doi.org/10.1016/j.chemosphere.2022.133577

37. Z. Ezaz, R. Azhar, A. Rana, S. Ashraf, M. Farid, A. Mansha, S.A.R. Naqvi, A.F. Zahoor, N. Rasool, Current trends of phytoremediation in wetlands: mechanisms and applications, in *Plant Ecophysiology and Adaptation Under Climate Change: Mechanisms and Perspectives II* (Springer Singapore, 2020), pp. 747–765. https://doi.org/10.1007/978-981-15-2172-0_28

38. R. Kumar, A. Mishra, M.K. Goyal, Water neutrality: concept, challenges, policies, and recommendations. Groundw. Sustain. Dev. **26**, 101306 (2024). https://doi.org/10.1016/j.gsd.2024.101306

39. M.K. Goyal, S. Rakkasagi, S. Shaga, T.C. Zhang, R.Y. Surampalli, S. Dubey, Spatiotemporal-based automated inundation mapping of Ramsar wetlands using Google Earth Engine. Sci. Rep. **13**(1), 17324 (2023). https://doi.org/10.1038/s41598-023-43910-4

40. I.R. Orimoloye, S.P. Mazinyo, A.M. Kalumba, W. Nel, A.I. Adigun, O.O. Ololade, Wetland shift monitoring using remote sensing and GIS techniques: landscape dynamics and its implications on Isimangaliso Wetland Park, South Africa. Earth Sci. Inf. **12**, 553–563 (2019)

41. R. Bi, C. Zhou, Y. Jia, S. Wang, P. Li, E.S. Reichwaldt, W. Liu, Giving waterbodies the treatment they need: a critical review of the application of constructed floating wetlands. J. Environ. Manag. **238**, 484–498 (2019). https://doi.org/10.1016/j.jenvman.2019.02.064

42. S. Pasika, S.T. Gandla, Smart water quality monitoring system with cost-effective using IoT. Heliyon **6**(7), e04096 (2020). https://doi.org/10.1016/j.heliyon.2020.e04096

43. M.K. Goyal, S. Kumar, A. Gupta, *AI for Water Treatment* (2024), pp. 31–40. https://doi.org/10.1007/978-3-031-72014-7_3

Chapter 5
Strategies for Wetland Ecosystem Restoration and Monitoring

Abstract In this chapter, we provide an overview of the findings from the assessment of wetland threats. It emphasizes the challenges posed by eutrophication and turbidity, specifically higher for inland wetlands impacted by human activities. Subsequently, we discuss the key issues impacting Ramsar wetlands, along with strategic policy and technological recommendations. It will assist in making policies aimed at restoring the wetland ecosystem and maintaining its health. The combined approach, leveraging both traditional knowledge and modern technologies to strengthen wetland conservation efforts, needs to be utilized. Considering this with a call to address existing research gaps, this section underscores the need for further investigation into less explored aspects of wetlands, fostering a deeper understanding of these ecosystems. This approach of integrating indigenous practices with technological advancements is essential for effective and sustainable wetland management. It enhances the conservation policies that balance ecological, social, and scientific perspectives. Lastly, we present insights into potential avenues for future research in wetland conservation.

Keywords Policies · Technologies · Community development · Future research

5.1 Summary of Findings

In this section, we summarize the discussion on the water quality, land cover patterns, and wetland health score assessment for all 85 Ramsar sites. The results were discussed, and major trends for different wetland types will be presented.

M. K. Goyal et al., *Monitoring India's Ramsar Wetlands*,
SpringerBriefs in Applied Sciences and Technology,
https://doi.org/10.1007/978-3-031-96818-1_5

5.1.1 Synthesis of Results from Coastal, Inland, and Human-Made Wetlands

All India Ramsar coastal, inland, and human-made wetlands are assessed, and their details are summarized as follows:

Coastal Wetlands

Coastal wetlands are substantially important wetlands with significantly larger disaster management responses. They consist of some of the biggest sites in the nation and thus support a wide variety of biodiversity. Mangrove vegetation across these sites is an important type of vegetation that helps to protect the shorelines from tidal effects [1]. For example, the Sunderbans and Bhitarkanika wetlands comprised significantly higher mangrove vegetation. Our study found that the major problems they face are human settlement and water quality modification. Eutrophication can also be seen in all the wetlands. The maximum eutrophication level can be found at the Vedanthangal wetlands, which is about 89.28%. Turbid areas can also be found in some places, but it isn't that concerning for these wetlands. The health card score prepared provided an idea about the vulnerability of the wetlands against the different types of threats. Among the coastal wetlands, Pallikarnai Marsh was the site with the most risk.

Inland and Human-Made Wetlands

These wetlands are situated within the periphery of the land and have great significance. Apart from supporting a significant amount of biodiversity, they serve other purposes like the freshwater supply, groundwater recharge, etc. For example, Yashwant Sagar Reservoir, a man-made wetland in Indore city, plays a major part in maintaining urban water, supplying water, and recharging groundwater in the surroundings. Any contamination of the wetland can cause great havoc in the surrounding populations. Among the examined wetlands, Longwood Shola Reserve Forest has a maximum eutrophication level with a value of 93.27%. Turbidity also seemed to be an issue in these wetlands. Around 12 sites have turbidity levels of more than 50%, with Keoladeo National Park seeming to have been worst affected, with 91.48%. Based on the wetland health card score, Deepor Beel is at the most risk, with a critically low score of 37.8 %. However, a few more sites are also observed on the verge of vulnerability.

5.1.2 Major Trends and Observations Across Different Wetland Types

Based on the assessment of all the sites, it seemed that the inland and coastal wetlands were equally affected by the eutrophication levels. Most of the coastal wetlands have observed eutrophicated areal percentage (based on NDCI) between 25 and 50, with

62.55% of sites falling under the category. In the case of inland wetlands, maximum wetlands have NDCI% between 0–25 at 39.13%, followed by NDCI% of 25–50 at 37.68%, which means they are comparatively less threatened by eutrophication as compared to the coastal wetlands, as shown in Fig. 5.1.

A similar analysis was done with the turbidity level of the sites in both categories. The coastal wetlands are in better condition in terms of turbidity, with all the sites having a turbidity area of less than 50% and most of them having even less than 25% turbidity area. The same is not true for the inland wetlands. While many of them have turbidity levels of less than 50%, few of them exceed that. Certain sites have turbidity levels of more than 75%, which is quite high. It depicts that inland wetlands face greater threats in terms of turbidity levels, as shown in Fig. 5.2.

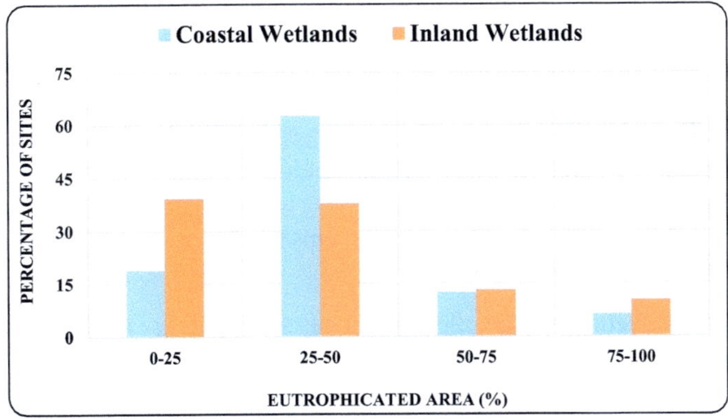

Fig. 5.1 Comparison of eutrophicated area percentage distribution

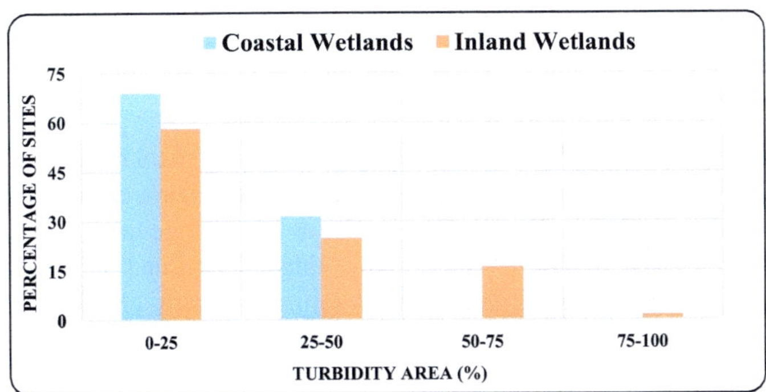

Fig. 5.2 Comparison of the turbidity area percentage distribution

5.2 Implications for Ramsar Wetlands

Several water quality parameters affect the ecosystem health of wetlands. Some examples are pH, dissolved oxygen, nutrient concentrations, and the content of contaminants. Depleting water quality from different causes, such as pollution, eutrophication, or hydrological changes, has negative impacts on wetlands. Some of the impacts on the Ramsar wetland sites are discussed in this section.

5.2.1 Impact of Water Quality on Ecosystem Health and Ramsar Site Management

The factors impacting the water quality, overall ecosystem health, and site management are discussed below:

Eutrophication: Over-enrichment of nutrients, especially those containing nitrogen and phosphorus, normally leads to eutrophication. Some eutrophication results in algae blooms that decrease oxygen, contributing to hypoxia or "dead zones" and thus to fish kills and disruptions in aquatic ecosystems by changes in species composition, affecting flora and fauna [2].

Contaminants: Wetlands also collect pollutants such as heavy metals, pesticides, or industrial waste. Contaminants interfere with species reproduction cycles, contribute to accumulation in the food chain, and sometimes make wetlands toxic, which could impair biodiversity along with water-dependent livelihoods [3].

Salinity Changes: Upstream pollution or altered hydrological regimes could significantly reduce the site's capacity to host its endemics due to possible variations in salinity in coastal Ramsar sites. Freshwater wetlands are gradually losing biodiversity as they become salty due to prevailing conditions that are not conducive to most aquatic plants and animals [4].

Changes in Hydrology: Wetlands require specific hydrological features to be met in terms of their ecological functions. Impoundments such as dams, canals, and even irrigation schemes alter inflow regimes into wetlands and change the seasonal flooding and drying cycles on which species rely for breeding, feeding, and shelter [5,6]. Indian Ramsar wetlands deteriorate significantly in terms of quality and thus degrade the site management aspects. Poor water quality directly affects the ecological character of the wetland; the Ramsar Convention treats this aspect very seriously as well. The recovery of water quality and fulfillment of the established Ramsar site conservation goals are the most important conditions for maintaining the balance in ecology and, thus, effective management of the site [7].

5.2.2 Recommendations for Policy and Management Practices

Effective policy and management measures will then be the determinant of the protection and restoration of health in Indian Ramsar wetlands. The main recommendations are as follows:

Water Quality Regulations: The government should impose stringent rules and regulations as per the present-day requirements on industrial discharge, untreated sewage, and agricultural runoff with substantial penalties. The Central Pollution Control Board (CPCB) and respective State Pollution Control Boards (SPCB) must observe water quality with more vigilance and implement formal standards. Regular audits of industries and effluent treatment plants near Ramsar wetlands should also be compulsory [8].

Community Involvement and Awareness: Engaging local communities is important in managing and conserving wetlands for the long run. Thus, increasing public participation, community-based water quality monitoring, and participatory approaches to management should be encouraged. Public awareness programs to reduce pollution, plastic waste, and overuse of water resources will foster a sense of responsibility toward wetland conservation [9].

Increase in Funds and Research: Beyond this, sufficient financial arrangements should also be made for the restoration and preservation of the Ramsar wetlands. Investment in scientific research to develop new technology for water purification, effective systems of pollution control, and biodiversity conservation measures is essential. The proposed measure also calls for the establishment of long-term monitoring programs so that the effectiveness of the pollution control programs can be monitored.

Integrated Watershed Management: Wetland management should be holistic and taken as a whole watershed. It would address pollution at its source, particularly agricultural runoff and municipal waste. Sustainable agriculture practices such as organic farming, reduction in chemical fertilizers and pesticide use, and the promotion of buffer zones of vegetation around wetlands may improve water quality significantly [10].

Wetland-Specific Management Plans: Every Ramsar site needs to have a site-specific management plan that responds to unique challenges. The plans must include measures to regulate the use of water, control invasive species, and control the hydrological regimes necessary for good wetland health. The recommendations proposed will eventually improve the protection accorded to Indian Ramsar wetlands so that their ecological and socio-economic benefits are sustained for posterity [11].

5.3 Technological and Policy Recommendations

In light of these ever-mounting environmental issues, especially concerning wetland management, there is a growing need for the introduction of newer technologies to make better monitoring and policy implementation feasible. Indian Ramsar wetlands are threatened by pollution, habitat degradation, and climate change. The challenge lies in striking a balance between modern technological solutions and sound policy frameworks to safeguard critical ecosystems. Below are a few technological recommendations and the corresponding policy measures that would enhance the conservation and management of wetlands, as shown in Fig. 5.3.

5.3.1 In-Situ Monitoring: Need for Real-Time Data Gathering as Well as Integration in Field Monitoring

Real-time monitoring is fundamental for wetland ecosystems because of their need to know their health status in accurate terms. Most traditional monitoring systems depend on the collection of data at certain intervals, hence failing to detect extreme environmental changes that eventually result in significant long-term consequences. Real-time monitoring provides continuous data on parameters of water quality, such as pH, turbidity, dissolved oxygen, nutrient levels, hydrology, and biodiversity, and thus increases the response capability of management.

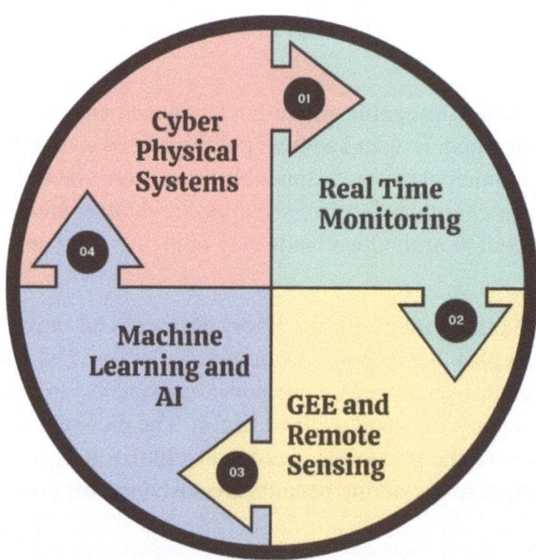

Fig. 5.3 Use of technology in wetland health monitoring

Importance of Continuous Data Collection: It would allow for the immediate detection of chemicals such as chemical runoff, heavy metals, or nutrient spikes that can produce harmful algal blooms, also known as eutrophication. In these conditions, immediate steps can be taken to prevent widespread ecological damage. Continuous monitoring of wetland health can provide a baseline understanding of environmental conditions and track changes due to natural events, such as monsoons or human activities.

Better Decision-Making: Sustaining the availability of data can improve the decision-making process for policymakers and conservationists, thereby enabling them to intervene in the form of pollution control measures, water regulation, or biodiversity conservation efforts much more promptly [12].

Integration with Field Monitoring: Although the fact is that real-time data collection indeed serves as a continuous, non-stop source of information, it is much more accurate and contextual if integrated with field-based monitoring. Field observations can check remote sensor data and identify species that sensors might overlook, bringing forth a better understanding of wetland conditions.

Policy Recommendations: A national guideline framework used for establishing real-time monitoring systems can foster cooperation between central, state, and local authorities. Implement real-time data monitoring for various industries around the Ramsar wetlands and make them adhere to environmental standards [13].

5.3.2 Google Earth Engine: Scaling Up Massive Environmental Observability

Google Earth Engine is a cloud-based tool that makes it possible to process a large amount of satellite imagery and geospatial data [14]. Being an ideal tool to monitor wetlands in large areas, GEE provides historical data over several decades to real-time data to allow changes in land use and water level, vegetation health, and habitat fragmentation within the Ramsar wetlands.

Advantages of GEE

Long-Term Trend Analysis: Long-term trend analysis of historical data can be provided, which can be useful in identifying long-term wetland health trends and impacts brought about by climate change, urbanization, or the expansion of agriculture.

Automated Change Detection: This can automatically identify changes in land cover, water bodies, and vegetation, which may provide valuable information for tracking unauthorized encroachments, deforestation, or water diversion within protected wetlands [15].

High Pace Data Processing: GEE's cloud platform helps to process big data within very short timescales, therefore offering virtually real-time insight into environmental changes.

Policy Recommendations: Adopt GEE at National Level: The government must encourage the implementation of GEE for wetland monitoring across every Ramsar wetland site so that the center can monitor the compliance of the environment through the environmental rules and regulations by implementing GEE into national and state-level frameworks of wetland management.

Capacity Building: Training programs should be established between government agencies, the local community, and conservation groups to strengthen their ability to utilize GEE and other remote sensing tools [16].

5.3.3 AI and ML: Integration into Water Quality Forecasting and Management

The development of Artificial Intelligence (AI) and Machine Learning (ML) will revolutionize the way environmental management is addressed through predictive analytics, automation, and good decision-making abilities. In the case of Ramsar wetlands, this may be applied through AI and ML models to predict changes in water quality, detect pollution patterns, and model the ecological impacts of different management interventions.

Application in Wetlands

- **Predictive Water Quality Models**: It enables AI-based algorithms with the capability to analyze historical and real-time water quality data and predict probable future changes based on the factors of seasonal variations, upstream pollution sources, and climate patterns. It helps in the anticipation of peaks in pollution levels and timely interventions.
- **Automatic Anomaly Detection**: Through ML algorithms, models can be trained to automatically identify water quality anomalies, including sudden pH level changes or any changes in dissolved oxygen levels could be indicative of a pollution event or ecosystem stress. Again, this avoids human error in interpretation.
- **Optimized Resource Use**: AI can be used to optimize resource utilization for wetland conservation; this could be the best time and location for interventions on pollution control or water releases in hydrologically stressed wetlands.
- **Policy Recommendations**
  1. Integration of AI in Environmental Policy: AI models in government water quality monitoring programs will include prediction tools the policymakers can use in managing and running the ecosystem.

2. The government should invest in AI and ML applications for wetland management by facilitating public–private partnerships toward innovation in that field [17].

5.3.4 Cyber-Physical Systems: Adoption for Effective Real-Time Monitoring and Rapid Response

Cyber-Physical Systems (CPS) couple the dynamics of physical processes with computation and networking in such a manner that it takes up intelligent systems to monitor and control the environment in real time. Integrating various monitoring technologies like sensors, drones, and satellite data with predictive models and response mechanisms, CPS can achieve a holistic approach to wetland management.

Advantages of CPS in Wetlands

- **Real-Time Response Capability**: Even with CPS, real-time automatic responses now allow for a reaction to environmental changes. For example, when water quality sensors in a system experience heightened pollution levels, they immediately alert the local authority or perform filtration systems.
- **Improved Monitoring Networks**: With CPS, one platform can merge data feeds from remote sensors, drones, and satellites so that the information can be analyzed and better-informed decision-making can be made. It allows for a more integrative view of the healthiness of wetland systems, thus allowing for proactive management.
- **Disaster Preparedness**: Based on CPS management of wetlands, one can predict the hazards that are destined for the wetlands in flood-prone or drought-prone areas [18]. Early systems tied to CPS can reduce the extent of damage to wetlands and affected communities.

Policy Recommendations

National CPS Initiatives: The government will develop national frameworks for deploying the systems in critical ecosystems, starting with the Ramsar wetlands, integrating the application of the CPS into existing frameworks of environmental monitoring systems, and promoting coordination between relevant stakeholders.

Inter-agency Collaboration: The application of Wetland Management through CPS will involve more than one sector working together. These sectors include environmental agencies, technology providers, and local governments. Thus, a policy that promotes collaboration will be vital to a successful application. It puts the country on the road to a better means of preservation and response as these technologies advance, such as real-time monitoring, Google Earth Engine, AI, and Cyber-Physical Systems integrated into wetland management. It leads the policymaking module to strive to promote the swift adoption of such technologies and their integration into national frameworks with a vibrant sense of the sustainability and health of the

Ramsar wetland ecosystem. In doing so, it safeguards those valuable ecosystems while fueling technological innovations in environmental management [19].

5.4 Community Development and Policy Making

Wetland ecosystems play an important role in supporting biodiversity and regulating water cycles, and they are crucial resources for local communities. Their conservation requires systematic coordination between different stakeholders, from policymakers to local communities. This section discusses the rise of the importance of local communities within the context of wetland conservation as well as an increase in community-based approaches through technology and strategies for enhanced collaboration between policymakers, scientists, and other stakeholders to manage wetlands better.

5.4.1 Role of Local Communities in Wetland Conservation and Management

Local communities are central to wetland conservation and management. Being close to the resource, they possess detailed knowledge of the ecosystems, what is going on in them, and where the challenges lie. Traditional Ecological Knowledge (TEK) is incredibly useful for evaluating early signs of environmental change, species interaction, and resource management. The knowledge gaps between scientific knowledge and on-the-ground realities could be bridged through the integration of TEK into traditional, formalized conservation practices. In most parts of the world, community-based activities such as the protection of waters through sustainable fishing techniques, traditional irrigation methods, and the securing of water bodies as religious sites have tremendously saved wetlands. The Indian Chilika Lake community has been engaged in resource management for fisheries and water quality monitoring, and this has effectively restored the wetland. Also, when communities are empowered to participate in decision-making processes, they are likely to adopt measures of conservation that both protect biodiversity and benefits derived from their livelihoods. Local communities also have economic and cultural ties to wetland ecosystems that make them integral stakeholders in any conservation effort [20]. Community-based supervision of wetlands is possible only when these communities are given proper knowledge, incentives, and resources. Inclusive policies that respect and integrate their rights and participation are vital for ensuring the long-term sustainability of wetland conservation programs.

5.4.2 Policy Recommendations for Integrating Technological Advancements with Community-Based Approaches

Technological advancement offers great opportunities for the improvement of wetland conservation when combined with community-based approaches. Some of the remote sensing technologies, like satellite imagery, Geo Information Systems, and GIS, will have real-time data on the health of a wetland, water quality, and biodiversity trends. It will necessitate the training of local communities to handle these tools to interpret and take action on the data generated to have proper resource management. Majorly, a policy recommendation is an investment in capacity-building measures that are geared toward equipping local stakeholders with skills to use modern technology at their helm effectively. For instance, environmental monitoring training programs on data collection, analysis, and monitoring through mobile applications or drones can be provided. Further, these technologies can be used by the community to monitor illegal activities like poaching or encroachment that may undermine the health of the wetland. Another measure that could be taken is the development of digital platforms for the sharing of information between communities, scientists, and policymakers. Some web-based platforms, like websites, mobile applications, etc., could make it easier for people to connect and get the required information. These could serve as repositories of traditional knowledge and environmental data in addition to best-practice conservation approaches. By integrating local insight with scientific data, policymakers can devise policies that are even better suited to the local context.

5.4.3 Strategies for Fostering Collaboration Between Policymakers, Scientists, and Local Stakeholders

The inclusion of all stakeholders in the management program is necessary to achieve the goal of conservation of wetlands. Collaboration between policymakers, scientists, and local stakeholders is key to addressing wetland issues. Some of the suggestions to enhance the collaboration are:

- **Establishment of Multi-stakeholder Platforms**: One strategy of institutionalizing effective management in such areas may include establishing multi-stakeholder platforms or forums that bring together all relevant parties to discuss issues, share knowledge, and work out joint plans for wetland conservation. It can be set at a level ranging from local to regional level, ensuring all voices, specifically those in the marginalized communities, are heard in decision-making.
- **Building Trust and Incentives for Collaboration**: Recognize and provide incentives for people working as a way of resolving the "implementation gap." Policymakers and scientists operate in silos, and very little overlap occurs. By entering into formal partnerships and bringing attention to the time and efforts necessary

to cross-gaps, collaboration is more rewarding and sustainable. Scientific institutions and local governing bodies should also be recognized and given rewards for valuing local knowledge efforts that are integrated into policymaking processes [13].

- **Participatory Governance Models**: These are important for the active involvement of local communities in policymaking through participatory governance models. In such models, policymakers and scientists interact with communities through regular consultations, participatory mapping exercises, and co-management agreements. India's NWCP is one example that promotes community participation in wetland management as a model of how inclusive governance can be put into practice in wetland conservation activities [21].

- **Long-Term Monitoring and Feedback Loops**: Long-term monitoring and feedback loops include continuing to collect data on key environmental and socio-economic indicators, thus allowing ecosystems to be adaptively managed. For wetlands, such as the Pantanal wetland, these indicators include biodiversity, water quality, and land use changes. Using the collected data, stakeholders such as planners and decision-makers can make adjustments to policies and conservation strategies as trends are observed. Feedback loops ensure that interventions respond to emergent challenges, thereby enhancing the health of the ecosystem while addressing sustainable development goals [22].

- **Educational Initiatives**: Increase awareness through education, focus on the value of wetlands and ecosystem services, and promote sustainable practices in communities and industries involved in the wetland region. It calls for policy frameworks to encourage people to cooperate [23]. Such policy frameworks may comprise incentives in the form of grants or subsidies to induce communities toward sustainable practices or adoption of wetland-monitoring technologies. Policies must be developed along such lines where the rights of indigenous peoples and local communities are advocated by ensuring that their knowledge is legally recognized and contributing to comprehensive national and regional conservation strategies.

5.5 Future Research Directions

The book provides a detailed assessment of wetlands, addressing their current health, threats, and their significance in sustaining biodiversity. It highlights the critical role wetlands play in preserving ecosystems, offering numerous ecological services in addition to habitat conservation. Further, there is a need for a more in-depth assessment to understand better their role in carbon sequestration and flood control, which are essential for both climate regulation and disaster management. Lastly, exploration of these benefits can enhance conservation efforts and help in developing strategies for sustainable wetland management.

References

1. M. Mishra, T. Acharyya, B. Halder, C.A.G. Santos, R.M. da Silva, N.R. Rout, D. Bhattacharyya, Impact assessment of Cyclone Yaas on the mangrove forest area in the Bhitarkanika National Park (India). J. Mar. Syst. **242**, 103947 (2024)
2. M. Devlin, J. Brodie, *Nutrients and Eutrophication* (2023), pp. 75–100. https://doi.org/10.1007/978-3-031-10127-4_4
3. C. Li, H. Wang, X. Liao, R. Xiao, K. Liu, J. Bai, B. Li, Q. He, Heavy metal pollution in coastal wetlands: a systematic review of studies globally over the past three decades. J. Hazard. Mater. **424**, 127312 (2022). https://doi.org/10.1016/j.jhazmat.2021.127312
4. G. Zhang, J. Bai, C.C. Tebbe, Q. Zhao, J. Jia, W. Wang, X. Wang, L. Yu, Salinity controls soil microbial community structure and function in coastal estuarine wetlands. Environ. Microbiol. **23**(2), 1020–1037 (2021). https://doi.org/10.1111/1462-2920.15281
5. M.K. Goyal, C.S.P. Ojha, D.H. Burn, Nonparametric statistical downscaling of temperature, precipitation, and evaporation in a semiarid region in India. J. Hydrol. Eng. 17(5), 615–627 (2012)
6. A. Azhoni, M.K. Goyal, Diagnosing climate change impacts and identifying adaptation strategies by involving key stakeholder organisations and farmers in Sikkim, India: challenges and opportunities. Sci. Total. Environ. **626**, 468–477 (2018)
7. R. Kumar, H. Ganapathi, S. Palmate, *Wetlands and Water Management: Finding a Common Ground* (2021), pp. 105–129. https://doi.org/10.1007/978-981-16-1472-9_5
8. CGWA, *Water Audit Notification* (2020), https://www.mpcb.gov.in/sites/default/files/water-quality/standards-protocols/Ground_Water_NewGuidelinesNotifiedeng24092021102020.pdf
9. S.J. Kakuba, J.M. Kanyamurwa, Management of wetlands and livelihood opportunities in Kinawataka wetland, Kampala-Uganda. Environ. Chall. **2**, 100021 (2021). https://doi.org/10.1016/j.envc.2020.100021
10. A.T. Hansen, T. Campbell, S.J. Cho, J.A. Czuba, B.J. Dalzell, C.L. Dolph, P.L. Hawthorne, S. Rabotyagov, Z. Lang, K. Kumarasamy, Integrated assessment modeling reveals near-channel management as cost-effective to improve water quality in agricultural watersheds. Proc. Natl. Acad. Sci. **118**(28), e2024912118 (2021)
11. J. Sucholas, Z. Molnár, Ł Łuczaj, P. Poschlod, Local traditional ecological knowledge about hay management practices in wetlands of the Biebrza Valley, Poland. J. Ethnobiol. Ethnomed. **18**(1), 9 (2022)
12. M.K. Goyal, S. Kumar, A. Gupta, *AI for Water Policy* (2024), pp. 41–53. https://doi.org/10.1007/978-3-031-72014-7_4
13. M.K. Goyal, S. Kumar, A. Gupta, *Basics of AI for Water Management* (2024), pp. 1–16. https://doi.org/10.1007/978-3-031-72014-7_1
14. S. Dubey, H. Gupta, M.K. Goyal, N. Joshi, Evaluation of precipitation datasets available on Google earth engine over India. Int. J. Climatol. **41**(10), 4844–4863 (2021)
15. I.R. Orimoloye, S.P. Mazinyo, A.M. Kalumba, W. Nel, A.I. Adigun, O.O. Ololade, Wetland shift monitoring using remote sensing and GIS techniques: landscape dynamics and its implications on Isimangaliso Wetland Park, South Africa. Earth Sci. Inf. **12**, 553–563 (2019)
16. M.K. Goyal, S. Rakkasagi, S. Shaga, T.C. Zhang, R.Y. Surampalli, S. Dubey, Spatiotemporal-based automated inundation mapping of Ramsar wetlands using Google Earth Engine. Sci. Rep. **13**(1), 17324 (2023). https://doi.org/10.1038/s41598-023-43910-4
17. R. Kumar, M.K. Goyal, Role of AI&ML in modernizing water and wastewater treatment processes. Water Air Soil Pollut. **236**(1), 31 (2025). https://doi.org/10.1007/s11270-024-07618-z
18. M. Kumar Goyal, V. Poonia, V. Jain, Three decadal urban drought variability risk assessment for Indian smart cities. J. Hydrol. **625**, 130056 (2023)
19. C. Alexandra, K.A. Daniell, J. Guillaume, C. Saraswat, H.R. Feldman, Cyber-physical systems in water management and governance. Curr. Opin. Environ. Sustain. **62**, 101290 (2023). https://doi.org/10.1016/j.cosust.2023.101290

20. UNESCO, *Chilika Lake* (World Heritage Convention, 2024), https://whc.unesco.org/en/tentat ivelists/5896/

21. V. Jain, K.S. Rautela, M.K. Goyal, Ecological restoration: an overview of science and policy regime, in *Ecosystem Restoration: Towards Sustainability and Resilient Development* (2023), pp. 1–27

22. K.M. Wantzen, M.L. Assine, I.M. Bortolotto, D.F. Calheiros, Z. Campos, A.C. Catella, R.M. Chiaravalotti, W. Collischonn, E.G. Couto, C.N. da Cunha, G.A. Damasceno-Junior, C.J. da Silva, A. Eberhard, A. Ebert, D.M. de Figueiredo, M. Friedlander, L.C. Garcia, P. Girard, S.K. Hamilton, C. Urbanetz, The end of an entire biome? World's largest wetland, the Pantanal, is menaced by the Hidrovia project which is uncertain to sustainably support large-scale navigation. Sci. Total. Environ. **908**, 167751 (2024). https://doi.org/10.1016/j.scitotenv.2023. 167751

23. U. Pasquier, R. Few, M.C. Goulden, S. Hooton, Y. He, K.M. Hiscock, "We can't do it on our own!"—integrating stakeholder and scientific knowledge of future flood risk to inform climate change adaptation planning in a coastal region. Environ. Sci. Policy **103**, 50–57 (2020). https:// doi.org/10.1016/j.envsci.2019.10.016

Correction to: Monitoring India's Ramsar Wetlands

Correction to:
M. K. Goyal et al., *Monitoring India's Ramsar Wetlands*,
SpringerBriefs in Applied Sciences and Technology,
https://doi.org/10.1007/978-3-031-96818-1

The original version of this book was inadvertently published with an inappropriate image for Figure 2.1 and 3.1 in Chapter 2 and 3, respectively, which has now been updated with a correct image in page 18 and 30. The book and the chapters have been updated with the changes.

The updated version of these chapters can be found at
https://doi.org/10.1007/978-3-031-96818-1_2
https://doi.org/10.1007/978-3-031-96818-1_3

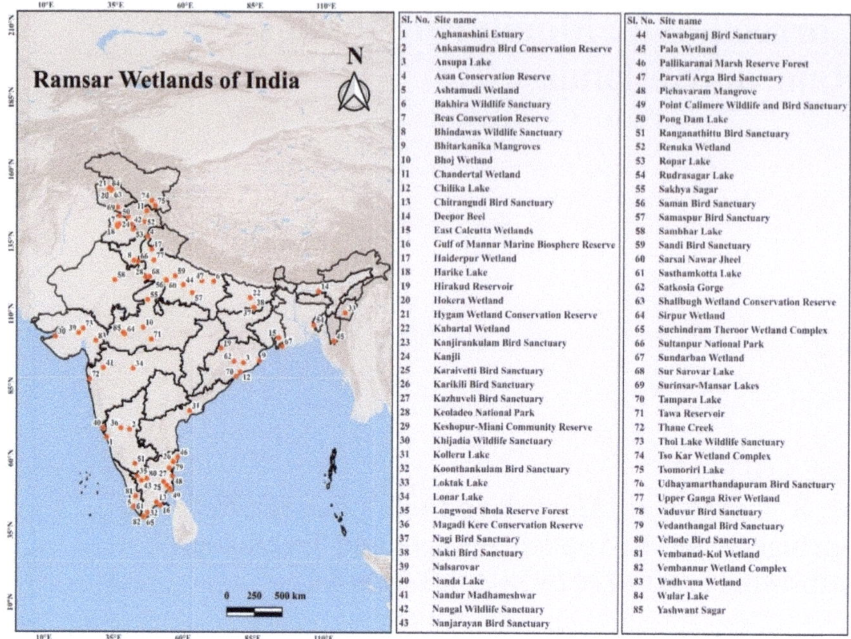

Sl. No.	Site name	Sl. No.	Site name
1	Aghanashini Estuary	44	Nawabganj Bird Sanctuary
2	Ankasamudra Bird Conservation Reserve	45	Pala Wetland
3	Ansupa Lake	46	Pallikaranai Marsh Reserve Forest
4	Asan Conservation Reserve	47	Parvati Arga Bird Sanctuary
5	Ashtamudi Wetland	48	Pichavaram Mangrove
6	Bakhira Wildlife Sanctuary	49	Point Calimere Wildlife and Bird Sanctuary
7	Beas Conservation Reserve	50	Pong Dam Lake
8	Bhindawas Wildlife Sanctuary	51	Ranganathittu Bird Sanctuary
9	Bhitarkanika Mangroves	52	Renuka Wetland
10	Bhoj Wetland	53	Ropar Lake
11	Chandertal Wetland	54	Rudrasagar Lake
12	Chilika Lake	55	Sakhya Sagar
13	Chitrangudi Bird Sanctuary	56	Saman Bird Sanctuary
14	Deepor Beel	57	Samaspur Bird Sanctuary
15	East Calcutta Wetlands	58	Sambhar Lake
16	Gulf of Mannar Marine Biosphere Reserve	59	Sandi Bird Sanctuary
17	Haiderpur Wetland	60	Sarsai Nawar Jheel
18	Harike Lake	61	Sasthamkotta Lake
19	Hirakud Reservoir	62	Satkosia Gorge
20	Hokera Wetland	63	Shallbugh Wetland Conservation Reserve
21	Hygam Wetland Conservation Reserve	64	Sirpur Wetland
22	Kabartal Wetland	65	Suchindram Theroor Wetland Complex
23	Kanjirankulam Bird Sanctuary	66	Sultanpur National Park
24	Kanjli	67	Sundarban Wetland
25	Karaivetti Bird Sanctuary	68	Sur Sarovar Lake
26	Karikili Bird Sanctuary	69	Surinsar-Mansar Lakes
27	Kazhuveli Bird Sanctuary	70	Tampara Lake
28	Keoladeo National Park	71	Tawa Reservoir
29	Keshopur-Miani Community Reserve	72	Thane Creek
30	Khijadia Wildlife Sanctuary	73	Thol Lake Wildlife Sanctuary
31	Kolleru Lake	74	Tso Kar Wetland Complex
32	Koonthankulam Bird Sanctuary	75	Tsomoriri Lake
33	Loktak Lake	76	Udhayamarthandapuram Bird Sanctuary
34	Lonar Lake	77	Upper Ganga River Wetland
35	Longwood Shola Reserve Forest	78	Vaduvur Bird Sanctuary
36	Magadi Kere Conservation Reserve	79	Vedanthangal Bird Sanctuary
37	Nagi Bird Sanctuary	80	Vellode Bird Sanctuary
38	Nakti Bird Sanctuary	81	Vembanad-Kol Wetland
39	Nalsarovar	82	Vembannur Wetland Complex
40	Nanda Lake	83	Wadhvana Wetland
41	Nandur Madhameshwar	84	Wular Lake
42	Nangal Wildlife Sanctuary	85	Yashwant Sagar
43	Nanjarayan Bird Sanctuary		

Fig. 2.1 Location map of Ramsar wetlands of India

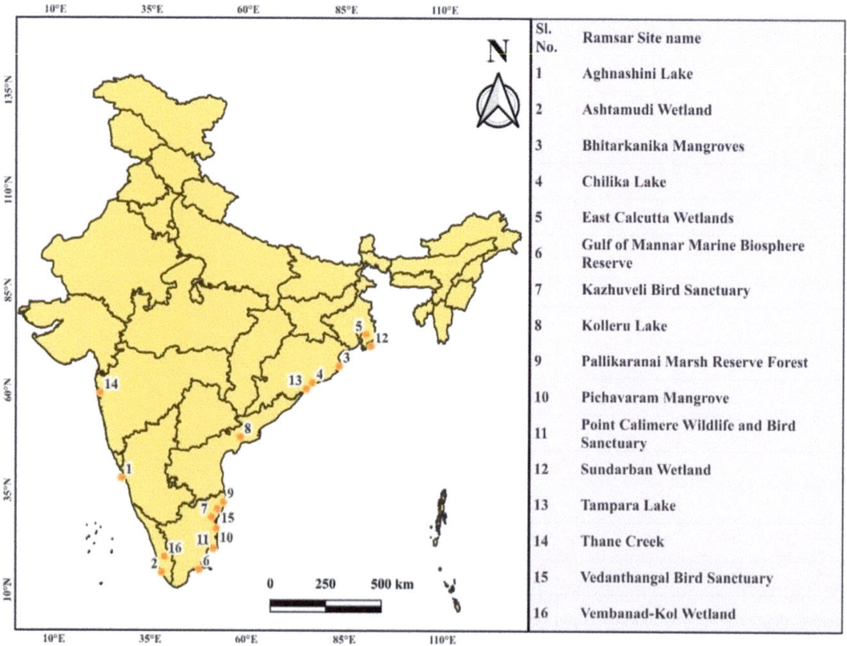

Sl. No.	Ramsar Site name
1	Aghnashini Lake
2	Ashtamudi Wetland
3	Bhitarkanika Mangroves
4	Chilika Lake
5	East Calcutta Wetlands
6	Gulf of Mannar Marine Biosphere Reserve
7	Kazhuveli Bird Sanctuary
8	Kolleru Lake
9	Pallikaranai Marsh Reserve Forest
10	Pichavaram Mangrove
11	Point Calimere Wildlife and Bird Sanctuary
12	Sundarban Wetland
13	Tampara Lake
14	Thane Creek
15	Vedanthangal Bird Sanctuary
16	Vembanad-Kol Wetland

Fig. 3.1 Location map of coastal wetlands of India